RESERVOIR ENGINEERING HANDBOOK

RESERVOIR ENGINEERING HANDBOOK

By

AMAR KUMAR

2012

SBS Publishers & Distributors Pvt. Ltd.
New Delhi

ISBN 13 : 9789380090535

First Published in 2012

Published by:

SBS PUBLISHERS & DISTRIBUTORS PVT. LTD.
2/9, Ground Floor, Ansari Road, Darya Ganj,
New Delhi - 110002,
INDIA
Tel: 0091.11.23289119 / 41563911 / 32945311
Email: mail@sbspublishers.com
www.sbspublishers.com

Printed in India by Chaman Enterprises, New Delhi.

Preface

Reservoir engineering involves more than applied reservoir mechanics. The objective of engineering is optimization. To obtain optimum profit from a field the engineer or the engineering team must identify and define all individual reservoirs and their physical properties, deduce each reservoir's performance, prevent drilling of unnecessary wells, initiate operating controls at the proper time, and consider all important economic factors, including income taxes. Early and accurate identification and definition of the reservoir system is essential to effective engineering. Conventional geologic techniques seldom provide sufficient data to identify and define each individual reservoir; the engineer must supplement the geologic study with engineering data and tests to provide the necessary information. Reservoir engineering is difficult.

Reservoir engineering has advanced rapidly during the last decade. The industry is drilling wells on wider spacing, unitizing earlier, and recovering a greater percentage of the oil in place. Techniques are better, tools are better, and background knowledge of reservoir conditions has been greatly improved. In spite of these general advances, many reservoirs are being developed in an inefficient manner, vital engineering considerations often are neglected or ignored, and individual engineering efforts often are inferior to those of a decade ago. Reservoir engineers often disagree in their interpretation of a reservoir's performance. It is not uncommon for two engineers to take exactly opposite positions before a state commission.

Author

Preface

Content

1

Introduction

EARTH'S WATER

Fig. 1.1 Emerald Bay

If people had to pick their favourite water body, they'd probably choose a crystal-clear lake nestled in the mountains. Not all lakes are clear or are near mountains, though. The world is full of lakes of all types and sizes. A lake really is just another component of Earth's surface water. A lake is where surface-water runoff have accumulated in a low spot, relative to the surrounding countryside.

It's not that the water that forms lakes get trapped, but that the water entering a lake comes in faster than it can escape, either via outflow in a river, seepage into the ground, or by evaporation. A reservoir is the same thing as a lake in many peoples' minds. But, in fact, a reservoir is a manmade lake that is created when a dam is built on a river. River water backs up behind the dam creating a reservoir. Here's a question for you: when a beaver dams a creek, is the pond that it creates a lake or a reservoir?

Fig. 1.2 Small Watershed Dam, Iowa.

The Earth has a tremendous variety of freshwater lakes, from fishing ponds to Lake Superior (the world's largest), to many reservoirs. Most lakes contain fresh water, but some, especially those where water cannot escape via a river, can be salty. In fact, some lakes, such as the Great Salt Lake, are saltier than the oceans. Most lakes support a lot of aquatic life, but the Dead Sea isn't called "Dead" for nothing—it is too salty for aquatic life! Lakes formed by the erosive force of ancient glaciers, such as the Great Lakes, can be thousands of feet deep.

Some very large lakes may be only a few dozen feet deep—Lake Pontchartrain in Louisiana has a maximum depth of only about 15 feet. Some of the salty lakes were formed in ancient times when they were connected to seas and when rainfall may have been heavier. These lakes have been shrinking since the last ice age. The ancient Lake Bonneville in the United States was once as big as Lake Michigan, and the Great Salt Lake was once about 14 times as large as it is now. Lakes are highly valued for their recreational, aesthetic and scenic qualities, and the water they contain is one of the most treasured of our natural resources.

Lakes constitute important habitats and food resources for a diverse array of fish, aquatic life, and wildlife, but lake ecosystems are fragile. Lake ecosystems can undergo rapid environmental changes, often leading to significant declines in their aesthetic, recreational, and aquatic ecosystem functions. Exposed to external effects from the atmosphere, their watersheds, and ground water, lakes are subject to change through time. Human activities can further accelerate the rates of change. If the causes of the changes are known, however, human intervention (lake-management practices) sometimes can control, or even reverse, detrimental changes.

LIMNOLOGY–THE STUDY OF LAKES

Limnology (the study of lakes and other freshwater systems) is the science that can provide improved understanding of lake ecosystem dynamics and information that can lead to sound management policies. As more studies are conducted on a variety of lake systems, the accumulated information leads to the development of general concepts about how lakes function and respond to environmental changes.

The condition of a lake at a given time is the result of the interaction of many factors—its watershed, climate, geology, human influence, and characteristics of the lake itself. With constantly expanding databases and increased knowledge, limnologists and hydrologists are able to better understand problems that develop in particular lakes, and further develop comprehensive models that can be used to predict how lakes might change in the future.

While the development of a limnological database and knowledge is important, no amount of generalization can provide a full understanding or predict conditions of any particular lake. Each lake system is unique, and its dynamics can be understood only to a limited degree based on information from other lakes.

Just as a physician would not diagnose an individual's medical condition or prescribe treatment without a personal medical examination, a limnologist or hydrologist cannot accurately assess a lake system or suggest a management strategy without data and analysis from that particular lake and its environment.

CHARACTERISTICS OF LAKES

The following are some of the most important basic factors that give unique character to each lake ecosystem.

Climate

Temperature, wind, precipitation, and solar radiation all critically affect the lake's hydrologic and chemical characteristics, and indirectly affect the composition of the biological community.

Precipitation is the main factor affecting run-off and the delivery of nutrients and sediments. Temperature, wind, and energy from the sun affect lake stratification and mixing, plant growth, and evaporation.

Atmospheric Inputs

The surface of a lake is directly exposed to atmospheric inputs. Not only wet precipitation, but also dry particles, can be major sources of certain contaminants to a lake. Each lake also receives indirect atmospheric inputs by way of the run-off from its watershed.

Geologic Substrate and Soils in the Basin

The soil type affects the potential for run-off and erosion. The physical characteristics of the substrate determine the extent, nature, and quality of ground-water inflows and outflows. These are primary factors affecting the lake's chemistry, because of transfers between water and sediments, and input of sediment, minerals, and nutrients from the watershed by run-off water flowing into the lake.

Physiography

The area, surface topography, existence of upstream lakes and wetlands, altitude, and land slope of the lake's watershed affect surface-water run-off and the amount and nature of chemicals and sediments entering the lake. The physiography of the region affects the size of a lake's watershed and ground-water contributing area. The boundaries of a lake watershed and ground-water contributing area may not necessarily coincide. Interactions with land use by people can appreciably change how these factors affect run-off and the export of nutrients and sediment.

Land Use

The type, location, extent, and history of land cover/land use (such as agriculture, rural, and urban developed areas) can greatly affect the quantity of surfacewater and ground-water inflows and outflows, as well as the amounts and types of sediment, nutrients and chemicals (natural or synthetic) that are transported into the lake from the watershed.

Lake Morphometry

Size, shape, and depth characteristics of a lake are critical in determining currents and mixing of the lake, as well as its thermal and chemical stratification characteristics.

COMMON ENVIRONMENTAL PROBLEMS IN LAKES AND PROBABLE CAUSES

Eutrophication is the natural process of physical, chemical, and biological changes ("aging") associated with nutrient, organic matter, and silt enrichment of a lake. If the natural process is accelerated by human influences, it is termed "cultural" eutrophication.

Lakes are subject to a variety of physical, chemical, and biological problems that can diminish their aesthetic beauty, recreational value, water quality, and habitat suitability. Among the most common lake problems, and the conditions that often occur with eutrophication are the following.

Algal Blooms

Extensive and rapid growth of planktonic (floating and suspended) algae, caused by an increased input of nutrients (primarily phosphorus, but occasionally can also be caused by nitrogen), is a common problem in lakes. Lakes normally undergo aging over timescales of centuries or thousands of years, but the process can be accelerated rapidly to only decades by human activities that cause increases in sedimentation and nutrient inflow to the lake.

Accelerated eutrophication and excessive algal growth reduces water clarity, inhibits growth of other plants, and can lead to extensive oxygen depletion, accumulation of unsightly and decaying organic matter, unpleasant odours, and fish kills.

Sedimentation/Turbidity

Increases in accumulation and/or resuspension of sediments can be a detriment to water quality and habitat for many aquatic species. Such events usually are caused by heavy rains that produce erosion and intense run-off, carrying heavy sediment loads into lakes. High winds, boating activity, and bottom-feeding fish, such as carp, may also resuspend bottom sediments and increase turbidity.

Oxygen Depletion

Decreases in dissolved oxygen to less than 3 mg/L (milligrams per liter) in the water can be harmful or lethal to many desirable species of aquatic life. The primary mechanism of oxygen loss is consumption by high rates of respiration and organic decomposition. Ideally, such consumption is offset by oxygen inputs from the atmosphere and from photosynthesis by aquatic plants.

However, in stratified lakes, the atmospheric source is cut off from the hypolimnion (deep lake layer), and oxygen concentrations in the hypolimnion may decline to zero (anoxia) until the lake mixes again. Under anoxic conditions, phosphorus may be released from the bottom sediments into the overlying water. This "internal loading" may be considerable with phosphorus-enriched sediments and prolonged anoxia. Prolonged low oxygen concentrations in the summer or under ice in the winter can lead to fish kills.

Growth of Aquatic Plants (Macrophytes)

Normal macrophytic growth generally is beneficial for the

lake ecosystem; among other benefits, the plants provide refuge for fish and other organisms. However, in some lakes, the growth of aquatic plants ("weeds") can become excessive and create a serious nuisance for lake users, interfering with swimming, boating, and other recreational activities.

Excessive macrophytes commonly are caused by increased nutrients, invasion of exotic species, or accumulation of organic sediment. The improvement of water clarity resulting from management actions designed to control algal production can provide better conditions for growth of rooted plants.

Water-level Changes

Wide fluctuations in stage (lake level) can create major hardships for lakeside residences, marinas, and businesses, and they also may impair the habitat suitability for nearshore biota. These changes most commonly are linked to weather anomalies (extended periods of abnormally high or low precipitation), but also may be associated with human activities such as withdrawals for water use.

Species Shifts

Populations of desirable animal and plant species might decline sharply or disappear, to be replaced by other species. Usually, the new dominant species will become a nuisance and degrade some or all desirable qualities of the lake. Species shifts can be caused by introduction of invasive species that may have little or no natural controls on their population growth, or are stimulated by changes in environmental conditions (for example, climate changes, acidification from "acid rain" or other changes in water chemistry, or physical changes).

WHAT OIL IS AND WHERE IT COMES FROM

- According to the most widely accepted theory, oil is composed of compressed hydrocarbons, and was formed millions of years ago in a process that began when aquatic plant and animal remains were covered by layers of sediment—particles of rock and mineral.

Over millions of years of extreme pressure and high temperatures, these particles became the mix of liquid hydrocarbons that we know as oil. Different mixes of plant and animal remains, as well as pressure, heat, and time, have caused hydrocarbons to appear today in a variety of forms: crude oil, a liquid; natural gas, a gas; and coal, a solid. Even diamonds are a form of hydrocarbons.

- The word "petroleum" comes from the Latin words *petra*, or rock, and *oleum*, oil. Oil is found in reservoirs in sedimentary rock. Tiny pores in the rock allowed the petroleum to seep in. These "reservoir rocks" hold the oil like a sponge, confined by other, non-porous layers that form a "trap."
- The world consists of many regions with different geological features formed as the Earth's crust shifted. Some of these regions have more and larger petroleum traps. In some reservoir rock, the oil is more concentrated in pools, making it easier to extract, while in other reservoirs it is diffused throughout the rock.
- The Middle East is a region that exhibits both favourable characteristics—the petroleum traps are large and numerous, and the reservoir rock holds the oil in substantial pools. This region's dominance in world oil supply is the clear result. Other regions, however, also have large oil deposits, even if the oil is more difficult to identify and more expensive to produce.

DRILLING FOR OIL

- To identify a prospective site for oil production, companies use a variety of techniques, including core sampling—physically removing and testing a cross section of the rock—and seismic testing, where the return vibrations from a man-made shockwave are measured and calibrated. Advances in technology have made huge improvements in seismic testing.

- After these exploratory tests, companies must then drill to confirm the presence of oil or gas. A "dry hole" is an unsuccessful well, one where the drilling did not find oil or gas, or not enough to be economically worth producing. A successful well may contain either oil or gas, and often both, because the gas is dissolved in the oil. When gas is present in oil, it is extracted from the liquid at the surface in a process separate from oil production.
- Historically, drilling a "wildcat" well—searching for oil in a field where it had not yet been discovered—had a low chance of success. Only one out of five wildcat wells found oil or gas. The rest were dry holes. Better information, especially from seismic technology, has improved the success rate to one out of three and, according to some, one in two. Reducing the money wasted on dry holes is one of the aspects of upstream activity that has allowed the industry to find and produce oil at the prices prevailing over much of the 1990's.
- After a successful well identifies the presence of oil and/or gas, additional wells are drilled to test the production conditions and determine the boundaries of the reservoir. Finally, production, or "develo-pment," wells are put in place, along with tanks, pipelines and gas processing plants, so the oil can be produced, moved to markets and sold. Once extracted, the crude oil must be refined into usable products, as discussed in the chapter on oil refining.

HOW OIL IS PRODUCED

- The naturally occurring pressure in the underground reservoir is an important determinant of whether the reservoir is economically viable or not. The pressure varies with the characteristics of the trap, the reservoir rock and the production history. Most oil, initially, is produced by "natural lift" production methods: the pressure underground is high enough to force the oil to the surface. Reservoirs in the Middle East tend to be long-lived on "natural lift," that is, the reservoir pressure continues over time to be great enough to force the oil out. The underground pressure in older reservoirs, however, eventually dissipates, and oil no longer flows to the surface naturally. It must be pumped out by means of an "artificial lift"—a pump powered by gas or electricity. The majority of the oil reservoirs in the United States are produced using some kind of artificial lift.

- Over time, these "primary" production methods become ineffective, and continued production requires the use of additional "secondary" production methods. One common method uses water to displace oil, using a method called "waterflood," which forces the oil to the drilled shaft or "wellbore."
- Finally, producers may need to turn to "tertiary" or "enhanced" oil recovery methods. These techniques are often centered on increasing the oil's flow characteristics through the use of steam, carbon dioxide and other gases or chemicals. In the United States, primary production methods account for less than 40 per cent of the oil produced on a daily basis, secondary methods account for about half, and tertiary recovery the remaining 10 per cent.
- Both the varying reservoir characteristics and the physical characteristics of the crude oil are important components of the cost of producing oil. These costs can range from as little as $2 per barrel in the Middle East to more than $15 per barrel in some fields in the United States, including capital recovery. It is interesting to note that technological advances in finding and producing oil have made it possible to bring once-expensive deepwater Gulf of Mexico oil into production for less than $10 per barrel.

THE IMPACT OF UPSTREAM TECHNOLOGY

- Technology's contribution to finding oil is huge.

Technology cannot change geology but, by revolutionizing the information available about the features of a geologic structure, it has enhanced the likelihood of finding oil. A primary benefit is the ability to eliminate poor prospects, thus considerably reducing wasted expenditures on dry holes. In addition, drilling and production technologies have made it possible to exploit reservoirs that would formerly have been too costly to put into production and to increase the recovery from existing reservoirs.

- Technology also has contributed to making oil exploration and production safer for the industry and for the environment. Offshore production can be operated from onshore, with automatic shutoff systems to minimize the pollution risk. Infrared photography can pinpoint a trajectory of spilled oil, allowing equipment and personnel to be deployed quickly and effectively, thus minimizing damage.
- Most of the attention paid to keeping hydraulic systems running clean has focused on initial flushing followed by continuous and off-line filtration. Electrostatic filtering and centrifuge purification have emerged as more intense techniques that remove even the smallest particles, oxidation products, and water from hydraulic fluid. Particulate contamination itself is damaging enough. However, chemical activity between contaminants and the fluid tends to accelerate breakdown of the fluid by producing aging products. Aging products are contaminants that result from the chemical breakdown of fluid due to high temperatures, the presence of water and oxygen, and the deterioration of additives. This is why keeping fluid clean extends the life of fluid—it prevents the condition of the fluid from rapidly deteriorating.
- One aspect of maintaining fluid cleanliness has long been relatively ignored: reservoir design. Standardization has helped make hydraulic systems

easier to design and certainly more economical, but has caused designers to almost universally specify the conventional rectangular, flat-bottom reservoir, Figure 1.3. As will be demonstrated, the conventional reservoir not only falls far short of an optimum design, but actually can promote fluid contamination.

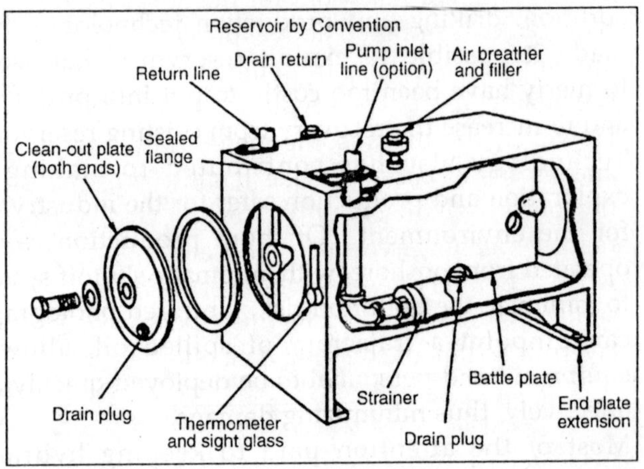

Fig. 1.3 Conventional Reservoirs Contain many Surfaces and Corners where Contaminants can Collect. Instead, Contaminants Should Remain Suspended in the Fluid for Easy Removal by Filters.

CONVENTIONAL RESERVOIRS AND THEIR SHORTCOMINGS

The reservoir is a basic part of any open hydraulic system and has considerable effect on the daily operation of the system. *The roles of the reservoir are to*:

- Accommodate the variations in the free fluid volume over a working cycle,
- Reduce turbulence, so air can escape from fluid and dirt can settle to the bottom of the tank,
- Serve as a mounting surface for pump, motor, valves, accessories, and other components.
- The usual configuration that meets these requirements

is a rectangular container with flat sides, ends, top, bottom and at least eight corners, Figure 1.3. Inside is a baffle, screen, or both to slow down the flow of fluid from its entry into the reservoir until it reaches the pump suction line inlet.

- The rectangular reservoir, for the most part, meets the requirements outlined above, provided that sediments are removed regularly from the reservoir floor and that the inside of the reservoir is cleaned periodically. However, the many components inside the reservoir and its many corners make it difficult to be cleaned thoroughly. Consequently, technicians may ignore regular cleaning schedules and attempt cleanup only when absolutely necessary. Unfortunately, a condition similar to that shown in Figure 1.4 often results—thick layers of sediment covering the bottom of the reservoir.

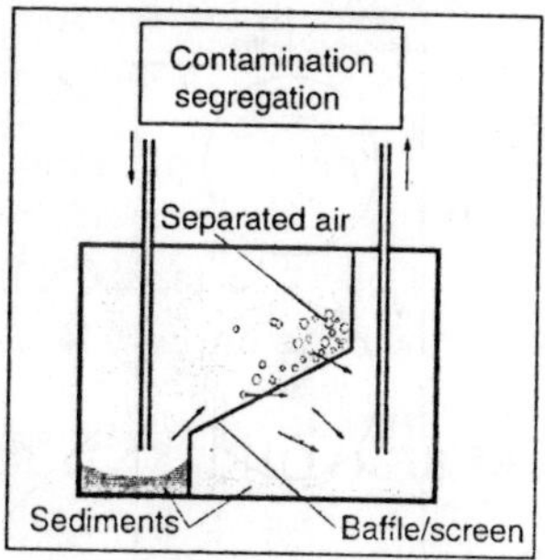

Fig. 1.4 Baffles and Screens Reduce Fluid Velocity, which Helps Remove air from the Fluid. Unfortunately, the Low Velocity Allows Contaminants to Settle, so they cannot be Removed by Filters.

- Realizing the severe negative effect on fluid cleanliness from the long term influence of settled contamination—combined with the low priority given to cleaning out the reservoir—perhaps some of the roles of the reservoir should be altered. For example, letting contaminants settle onto the bottom of the reservoir keeps them in continuous contact with the fluid, thereby affecting its chemistry. Moreover, the surface of this sediment layer is very unstable, so a change in the return oil flow rate or direction can stir up clouds of contamination in the otherwise clean fluid. Even with careful flow regulation, stirring up this sediment layer cannot be avoided in mobile and marine applications.

A NEW STRATEGY

Contaminants, therefore, should not be allowed to settle in the reservoir; they should be removed as soon as possible through filters and water coalescers. The most effective way to do this is to adopt a reservoir shape different from the traditional "bath tub" that has become so prevalent. This shape has already been in use for decades in aviation applications, where any disturbance in the operation of a hydraulic system could prove catastrophic.

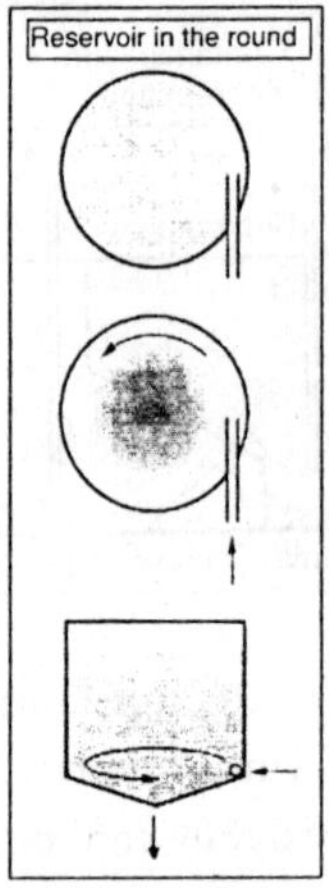

Fig. 1.5 Cylindrical Reservoir with Stationary Oil, Top, Leaves Contaminants Equally Distributed.

One self-cleaning reservoir is presented in Figure 1.5. The reservoir takes the form of a vertical cylinder with a conical bottom. A smooth transition joint exists where the vertical side wall meets the conical bottom. The suction connection for a filter or filtration circuit is located in the lowermost part of the reservoir, at the apex of the inverted cone. The return flow inlet is tangential to the cylindrical wall. When the return flow enters the reservoir, the entire mass of oil tends to slowly rotate.

Rotating the fluid, middle, directs contaminants to the center of rotation. As rotational velocity diminishes, contaminants settle at the bottom center of the reservoir, where they can be filtered out.

Rotating the mass of fluid directs contaminants towards the center of the reservoir. This action can simply be demonstrated by stirring a cup of tea. Fluid dynamics divert the tea leaves into the vortex of the spinning fluid. The same principles govern the rotating hydraulic fluid: solid contaminants concentrate at the center of rotation. This principle differs from that in a centrifuge, where rapid spinning introduces centrifugal force many times that of gravity, which directs particles away from the center of rotation. Unlike fluid in a centrifuge, fluid in the reservoir rotates relatively slowly, so fluid dynamics have a much greater effect on particles than centrifugal force does. Because the solids are more dense than the hydraulic oil, they eventually settle to the bottom center of the reservoir, where the filtration circuit inlet is located. From there, contaminants will be trapped by the filter instead of collecting at the bottom of a flat surface.

It is essential to rotate the mass of oil under normal operating conditions to keep particles, water, and other foreign contaminants from settling. In this manner, contamination is kept suspended in the oil until it is caught in a filter. Most of the contamination is taken out of circulation the first time it encounters a filter. Furthermore, filters are much easier to maintain than reservoirs are. Replacing a dirty filter element has become easier and more convenient than ever, and indicators make it easy to determine when an element needs to be changed. Cleaning reservoirs, on the other hand, is an unpleasant, labor-intensive procedure. Moreover, it is difficult to determine how dirty a conventional reservoir is without emptying and inspecting it. By the time you go through all the trouble, you should probably clean it anyway.

A bypass filtration circuit, sometimes called a kidney loop, works especially well with the cylindrical reservoir. A kidney loop often incorporates an oil cooler or heater as well. Fluid from the kidney loop flows back to the reservoir through a tangentially oriented tube, Figure 1.5. The main system pump draws fluid from the upper portion of the reservoir, where fluid is cleanest. Return oil flows tangentially into the reservoir near the bottom. No unwanted substances will be accumulated any place in the system because particles and other contaminants are continuously removed. At no time are they given an opportunity to accumulate anywhere. This results in simpler maintenance procedures, longer fluid life, and higher system reliability.

BETTER BY DESIGN

A secondary benefit of vertically oriented cylindrical reservoirs is better utilization of the materials than with conventional designs—especially when larger reservoirs are required. When the reservoir must exceed approximately one meter in height, hydrostatic pressure of the liquid column has to be considered to prevent the reservoir's side plates from deflecting outward. To reduce or prevent bulging and strain in the material, designers often specify corrugated side plates or heavy gauge steel for constructing the reservoir, or they specify ribs for added stiffness. Welded ribs, however, are expensive to install. Furthermore, locating them inside the reservoir creates corners and crevices that collect contamination. However, locating them outside the reservoir is a losing proposition too, because they consume additional floor space.

In a cylindrical reservoir, the hydrostatic pressure itself presents no practical impact on the thickness of the side plate. This is because the geometry of a cylinder distributes stress from the hydrostatic pressure evenly around the circumference of the reservoir. The construction need only be strong and stiff enough to allow transporting the empty tank, which serves as a base for its flanges and connections and carries the weight of the fluid. In comparing two reservoirs with identical volumes, a cubic reservoir with top, bottom, and sides 2 m square has a volume of 16 m3. A cylindrical reservoir with the same height and volume would have a diameter of about 2.55 m. With the cubic reservoir, side plates would have to be 14 mm thick to withstand the hydrostatic pressure from a column of oil, and bulging would be approximately 15 mm. On the other hand, plate thickness for the cylindrical reservoir would only have to be 0.25 mm or less. Furthermore, diametral deformation from hydrostatic pressure would be only about 1 mm.

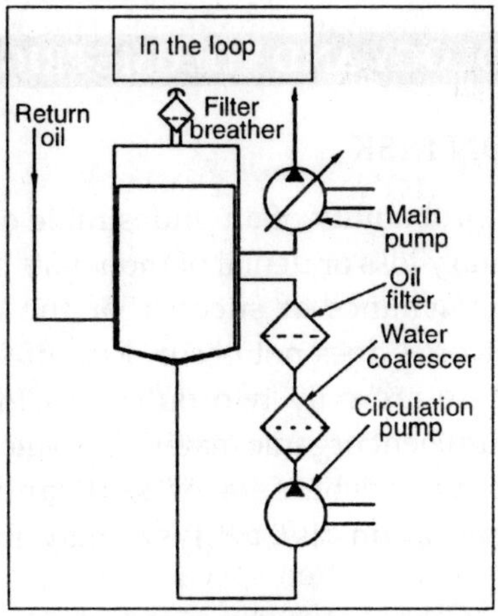

Fig. 1.6 System with Cylindrical Reservoir and off-line Filtration Circuit.

What this all boils down to is that cylindrical reservoirs not only promote cleaner fluid, but can be less expensive and lighter weight due to simpler design and more conservative material requirements.

2

Test Analysis Methods

BASIC GEOLOGY AND VOLUMETRIC ANALYSES

EXPLORATION RISK

Risk is the probability of an undesirable outcome, usually related to monetary loss or denial of monetary gain. Frequently, we talk in terms chance of success, or the chance that the undesirable outcome does not occur. Risk of success is 1- risk of failure, if there are only two outcomes. All hydrocarbon deposits need sufficient organic material, permeable formations, closure (trapping mechanism or seal) and time to form petroleum. Exploration risk analyses may include all these factors.

A final factor can be lumped into the risk of "non-commerciality." This factor can come from a host of concerns-excessive drilling cost, discovery of geologically complex or highly faulted/compartmentalized structure or a limited market for the products (gas, volatile oil, gas condensate, tar sands, kerogen). Risk factors are normally multiplied together to calculate the exploration risk, and the expected risk weighted return for investment. A newly discovered field that tests positive for hydrocarbons may be considered uneconomic to develop. The cash flow generated from the sale of products must be sufficient to justify the investment. Non-commercial wells are considered "dry holes" under US accounting procedures.

TRAPS

Traps may be structural, stratigraphic or a combination of these two. Growth (also normal faults) and reverse faults can be trapping mechanisms. Salt diapir (salt intrusion or salt piercement dome) is an effective trapping mechanism formed as the salt being of lighter density protrudes through layers above it. A pinchout is a form of stratigraphic trap. Also, an unconformity (missing time sequence) or an angular unconformity.

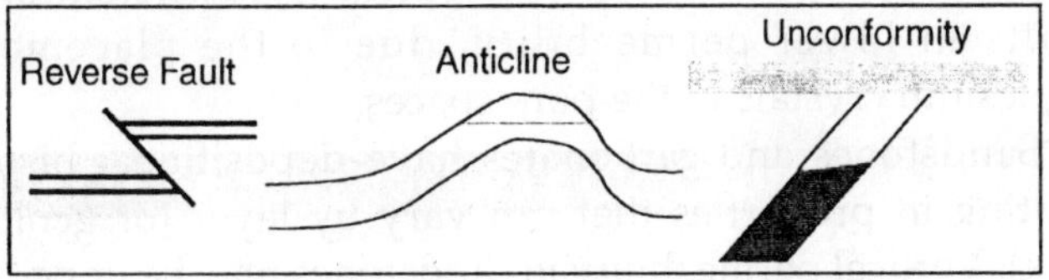

Fig. 2.1

Salt intrusion plugs are common in the Gulf of Mexico and other field in the world. The deposits directly above plugs can be highly faulted, making for very complex structures.

GEOLOGY AND ROCK PROPERTIES

Geologists and engineers work jointly to map out the stratigraphy of the deposits using fence diagrams. The initial evaluation stage involves intense evaluation of all petrophysical information. Commercial deposits are almost entirely either sandstone or carbonates at least at present (oil shale has a huge potential). Often sandstones are referred to as clastics, which means pieces of pre-existing rock. This makes carbonates, non-clastics. Pretty simple. One difference is that sandstone formations have depositional histories that involve transport of both organics and the sand particles. The barrier bar, deltaic marine deposits, and alluvial channel deposits all refer to sandstone deposits involving transport.

In contrast, carbonate formation developed more or less their present location. For oolitic limestone, oolites are calcareous precipitates that came from the sea water. In other cases, reefs and shoals grew and subsequently, marine life lived inside the structure ultimately becoming the organic material for hydrocarbon. The process by which deposited sediments undergo physical, chemical or biological change is referred to as diagenesis.

Generally, the process decreases porosity. However, dolomitization which is the dissolution of minerals in limestone results in improved porosity. The element magnesium replaces calcium, so the density of the dolomites is increased, in the range of 2.83 to 2.87 (grain density). In some cases, dolomitization results in lower permeability, due to the placement of magnesium crystals in the pore spaces.

Sandstones and carbonates have depositional histories, resulting in properties that can vary by layer (or geological strata). Vertical permeability in sandstones may be impeded by shale breaks while in carbonates, it may be clays or other minerals. In modeling fields, many feel that the vertical to horizontal permeability ratio is generally lower in sandstone.

Once the strata is identified in the logs (SP and Gamma Ray logs are used generally), a fence diagram is drawn between wells. Structural contour maps and isopach maps further define the deposits. As the field is delineated, further mapping of the field and associated layers are needed to identify petrophysical properties.

Additional Note- Tapping Unconventional Hydrocarbon Sources

Oil that has not matured sufficiently are called kerogens and the deposits referred to as oil shales, and require "cooking" or the process of pyrolysis to convert them into synthetic oils.. Another largely untapped resource are tar sands, which are biodegraded hydrocarbons. Coalbed methane is an unconventional gas source.

CALCULATION OF OIL AND GAS IN PLACE

Original oil in place is referred to the oil present at discovery. It may be abbreviated, OOIP, or OIP. It should always be referred to as the surface quantity in place. Similarly, original gas in place, can be either OGIP or GIP, and always referred to as the surface quantity in place.

$$N = 7758 \cdot A \cdot \phi \cdot h \cdot (1 - Swc) / Bo$$

$$G = 43560 \cdot A \cdot \phi \cdot h \cdot (1 - Swc) / Bg$$

Bo and Bg are Formation volume factors, and should always be expressed as reservoir volume to surface volume. Just remember it B is R before S. If there are 10 MSTB in the stock tank barrel, and Bo = 1.6, then there are 16 M rbbls were removed from the reservoir. A refers to the area of the structure in acres, porosity is expressed a fraction, and Swc is the initial or connate water saturation.

Bg and Bo can be found using correlations or from laboratory measurements. For Bg, the laboratory would identify the relationship of the Z-factor to the pressure at reservoir temperature, from which Bg can be calculated. If the Z-factor is considered to be 1, then Bg=1/p, where p is the reservoir pressure in psia.

The inverse of Bg is termed bg, and for z = 1 (ideal gas), bg = p, and this is a quick way to check if the Bg calculated is approximately correct. In calculating in-place volumes, cutoffs may be employed, based on minimum porosity, maximum water saturation and maximum volumes of clay or shale.

In these cases, it is most correct to calculate the porosity, thickness, hydrocarbon saturation product first for each log interval using the cutoffs, and then take an average of the product over the gross thickness of the oil column.

GAS PROPERTIES

Gas properties include density, formation volume factor, gas gradients, compressibility and viscosity. Additional properties are defined later, when discussing gas-oil systems and gas condensate systems. Viscosity of gas rarely measured in the laboratory. It is obtained from the Gas specific gravity is denoted as SG or γ_g and it is density of the gas at standard conditions (atmospheric pressure, 60 degrees F). Molecular weight of air is 28.97, so if the gas has a SG of 1.2, then the molecular weight would be 1.2*28.97.

Density of gas, ρ_g as presented below, is in lbs/cubic foot. For field units (degrees R, lb-mols, cu ft. and psia), the gas constant R = 10.73. As noted above, Bg is given in terms of reservoir volume to surface volume, so its units can be either Rcf/SCF or Rbbl.STB. Bg is generally less than 10^{-3} for 1,000 psi or higher reservoir pressure. Due to the awkwardness of carrying many decimal points, some textbooks use different units for Bg. Hopefully, newer textbooks will end his practice.

$$\rho_g = 2.7 \cdot \gamma_g \cdot p/(zT)$$

$$\rho_g = (p \cdot MW)/(10.73 \cdot T)$$

$$B_g = 0.0283 \cdot zT/p$$

$$PV = ZnRT, R = 10.73, n = \text{number of mols of gas}$$

The Z factor of gas is typically measured in the laboratory at reservoir temperature and for pressures ranging from atmospheric to reservoir pressure or above. Z factors can be calculated based on the composition of the gas, using the critical properties of components to calculate a pseudo-reduced pressure and temperature. Z- factors can be found in Craft and Hawkins. If only gas gravity is available, pseudo-critical properties are available.

Gas gradients are expressed in psi/ft or Grad =ρ_g/144 The gas gradient within a gas field are typically not significant. However, for flow up the wellbore, gas gradients will change, depending on temperature and pressures. Gas compressibility is defined below.

$$cg = \frac{1}{p} - \frac{1}{z}\left(\frac{dz}{dp}\right)$$

Generally, these values are included in laboratory tests, or they can be easily calculated from a known z-factor verses pressure curve. In a depleting oil reservoir, gas is liberated from the oil as the pressure declines below bubble point, and with any significant gas saturation in the oil, the system compressibility can rise dramatically:

$$c_t = c_r + S_w \cdot c_w + S_o \cdot c_o + S_g \cdot c_g$$

Water and oil compressibilities are on the order of 3 to 5E-6, while gas compressibility is in the 1E-3 to 1E-4 range. A 10% gas saturation can cause a 10 fold increase in system compressibility. The high compressibility of gas plays an important role in well testing and reservoir simulation. Rock compressibility can be obtained from Hall's correlation for consolidated rock.

ADDITIONAL NOTE: SOFTWARE USE IN RESERVOIR VISUALIZATION AND VOLUMETRICS

Commonly, automatic contouring programmes, cross-sections and 3-D visualization software combine to provide an extended view of the field. All attributes can be mapped, which includes oil gravity, oil viscosities, net-to-gross thicknesses, oil concentrations (h*por*So) or water saturations. As the field produces, contour maps can be made of reservoir pressures (isobaric maps), water-cuts and producing gas-oil ratios. Bubble maps of produced oil, water and gas are common.

Some subjectivity exists when contouring sparce data, particularly in the extension of contours beyond control points. In this case, geological principles, existing trends, and seismic (if mapping a surface) play a role in mapping. Software should be used in all cases to calculate reservoir volumes to remove the element of human error. In calculating reservoir volumc, older books refer to the trapezoidal formula and planimeter.

BASIC FLOW RELATIONS IN RESERVOIRS

PERMEABILITY AND RELATIVE PERMEABILITY

Movements of fluids in porous media are govern by three forces: viscous, capillary and gravitational. These forces depend on the properties of the porous media and fluids. At a microscopic level, fluids pass through irregular expanding and confining pathways. It would be a daunting task to calculate the frictional losses and compressional effects through the media at this level. Fortunately, on a lab scale to the field scale, Darcy's empirical equation holds and greatly simplifies the analysis. The single phase Darcy equation for steady state conditions is:

$$q = 0.001127 \frac{kA}{\mu B} \frac{\Delta p}{\Delta x}$$

where q is in units of STB/D and A is in units of square feet. Darcy's equation is a part of many of the flow analysis done by petroleum engineers.

Some of the areas where Darcy's equation plays a role are:

- Well test analysis
- Core flooding
- Fractional flow equation identifying displacement efficiencies
- Reservoir simulation

It is not used volumetic analysis and classical material balance equations. Some of the newer material balance programmes have added a deliverability equation, which relates to Darcy's equation.

MORE GENERAL FORM AS USED IN RESERVOIR SIMULATION

Darcy's empirical equation as used in formulation of simulation inter cell flow coefficients. The oil phase equation excluding gravity effects, is:

$$qo = -1.127 \times 10^{-3} \frac{k \cdot kro \cdot A}{\mu_o Bo} \frac{\partial p}{\partial x}$$

where q is in STB / D and A is in square ft.

The same equation can be written for water and gas phases in the reservoir. The Darcy equation can also be stated as:

$$q_F = -T\left(\frac{Kr}{\mu B}\right)_p \frac{\partial p}{\partial x} \text{ or } T\left(\frac{Kr}{\mu B}\right)_p \frac{\Delta p}{\Delta x}$$

$$\text{where } T = 1.127\times 10^{-3}\cdot KA$$

and q is in STB/D and A is in square ft.

where the subscript "p" stands for phase (oil, water and gas). Note that the relevant cross sectional area for flow in the x-direction is: $A = \Delta y \cdot \Delta z$.

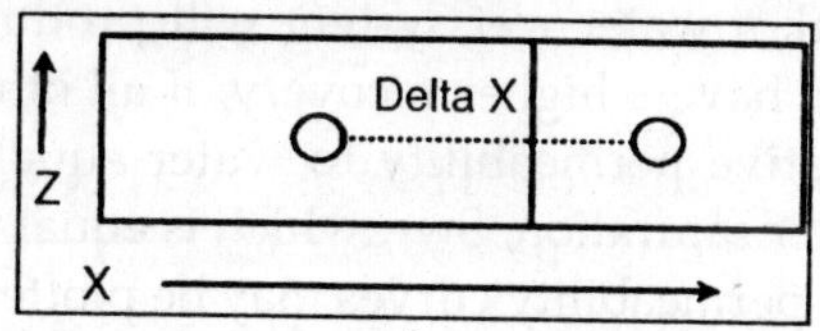

Fig. 2.2

$T_x = T/\Delta x$ where Δx is defined as distance between cell centers or $\Delta x = 0.5\Delta x_1 + 0.5\,\Delta x_2$

The term

$$\left(\frac{Kro}{\mu Bo}\right)$$

is a dynamic property and must be evaluated in every time step. Viscosity and Bo are calcuated based on pore volume-weighted average of the adjacent cell. Relative permeability is calculated based on the upstream saturation.

If the two cells are at the same level, then the upstream cell is simply the cell with higher pressure. However, if the cells are at different levels, and have the same pressure, the cell that has the higher elevation has a higher potential. Darcy's equation can be written to include the gravity term as follow:

$$q_F = -T\left(\frac{Kr}{\mu B}\right)_p\left[\frac{\partial p}{\partial x} - 0.433\gamma\cdot\sin\theta\right]$$

where γ is the specific gravity of the fluid

Simulation also includes the continuity equation, which is beyond the discussion, at least at this point.

OIL-WATER RELATIVE PERMEABILITY

Relative permeability is typically measured in the laboratory. A wetting phase is the phase that has higher surface tension than the other phases, so it spreads more over the internal pore surfaces. Natural formations are thought to be water wet, however due to oil migration, they can become oil wet. In general, a water-wet system will produce at a lower water cut, thus have a higher recovery, if all other factors are the same. Relative permeability to water equals zero at the maximum water saturation, Swr, which is equal to 1-Sor, or 1-Sorw. Relative permeability curves may be plotted on semi-log paper for easier comparisons between corefloods.

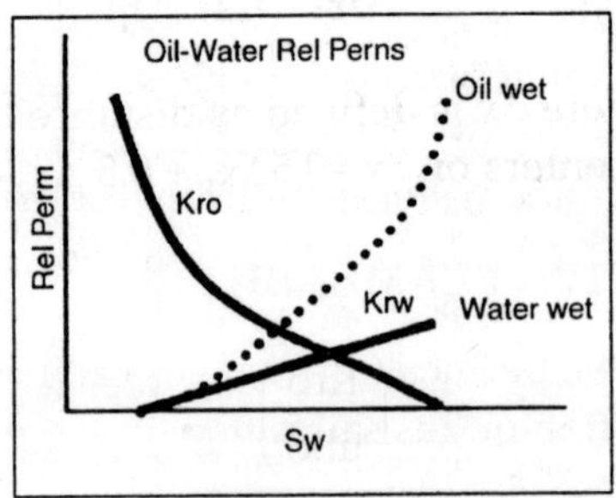

Fig. 2.3

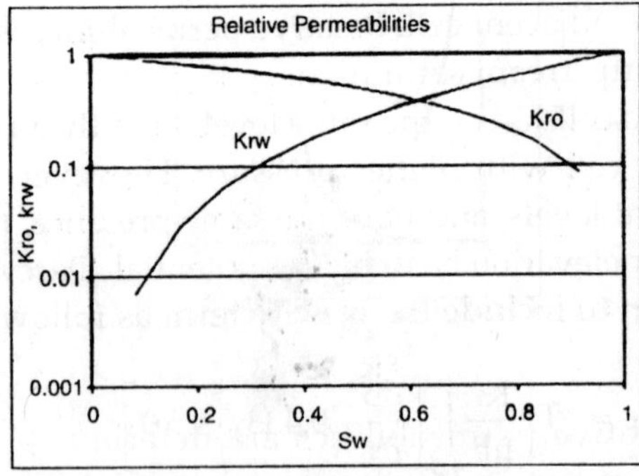

Fig. 2.4

Relative permeability curves may be plotted on semi-log paper (log scale on relative permeability scale) for easier comparisons between corefloods and extension of laboratory relative permeability data to saturation endpoints.

NORMALIZED CURVES

From core floods, relative permeabilities are determined from the minimum water saturation where only oil can flow to the maximum water saturation where only water flows. For purposes of averaging curves from multiple tests, the curves can be plotted on a normalized water saturation scale, where the satuation is equal to:

$$S_w^{\bullet} = \frac{S_w - S_{w,min}}{S_{w,max} - S_{w,min}}$$

Note that Sw,min is often referred to as the initial or connate water saturation. It is for the particular test, however due to limited coring, connate water is typically estimated from logs.

GAS-OIL RELATIVE PERMEABILITY

Gas-Oil relative permeability curves can be plotted verses total liquid saturation or gas saturation.

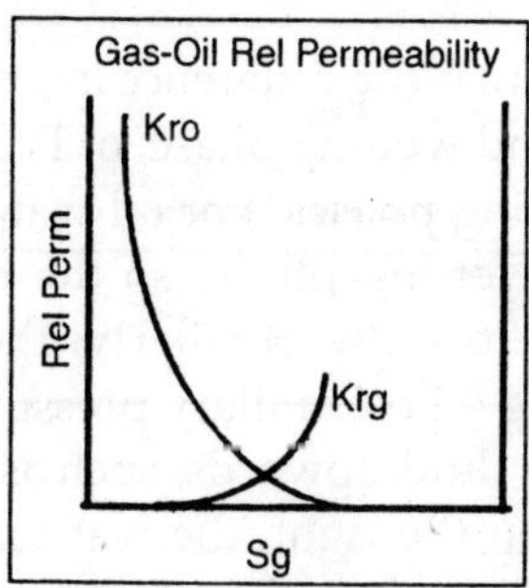

Fig. 2.5

When relative permeabilities are determined from core floods, values of Kro and Krg will be determined over a range of gas saturations (Sg, min to Sg, max) where there is two phase flow. The maximum gas saturation is related to the residual oil saturation to gas as follows:

$$S_l = 1 - S_g \text{ so } S_{l,min} = 1 - S_{l,min}$$

$$S_{l,min} = 1 - S_{wc} - S_{org}$$

For simulation, the gas-oil curves must be extended to relative permeability = 1, as follows:

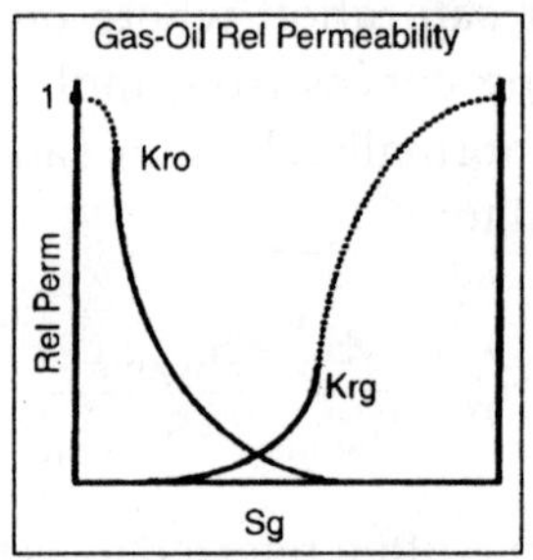

Fig. 2.6

Plotting the relative permeabilities on a semilog scale may improve the extrapolation of the curves.

HYSTERESIS EFFECTS ON CAPILLARY PRESSURE AND RELATIVE PERMEABILITY

Capillary pressure is the difference in pressure between the non-wetting phase and wetting phase, or Pcap = Pnw - Pw. Gas or air are never wetting phases. For oil-water system, water is assumed to be the wetting phase, so the oil-water capillary pressure is Pcow = Po – Pw. Similarly, the gas-oil capillary pressure is: Pcog = Pg – Po Capillary pressure is typically seen as a force that pulls a fluid upwards, such as absorbing fluid in a towel. As shown on the right, the water rises in a capillary tube. At the oil-water interface, the oil pressure equals the water pressure plus the capillary pressure. The porous media can be idealized as a bundle of capillary tubes. Under this analogy, capillary pressure is a function of saturation for a particular oil-water or gas-oil system. Capillary pressure is also a function of the direction in which saturation is changing. This phenomena, called hysteresis, means to correctly know the capillary pressure, we must know the history of saturation changes.

There are two cases:

1. *Drainage*: Flow resulting from a decrease in the wetting phase.
2. *Imbibition*: Flow resulting from an increase in the wetting phase.

The effect of hysteresis on oil-water capillary pressure is shown on the left. Typically, the drainage curve is used in simulation. This may seem incorrect as typically water entrusion (or water flooding) would increase Sw with time. However, capillary pressures are used to assign the correct initial water saturations within the reservoir. As the formation is "filled", oil displaces water, decreasing the wetting phase (water).

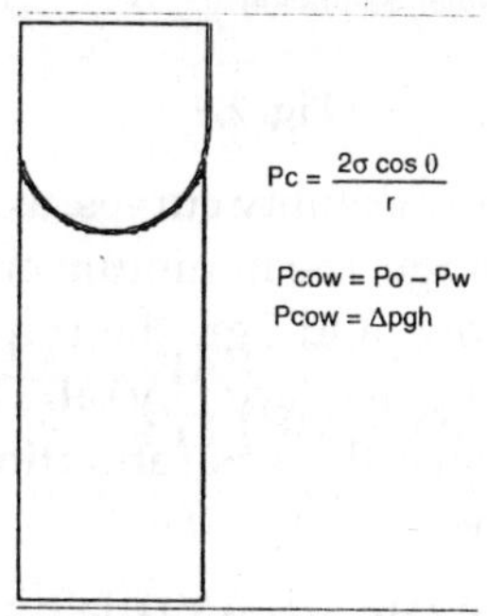

Fig. 2.7

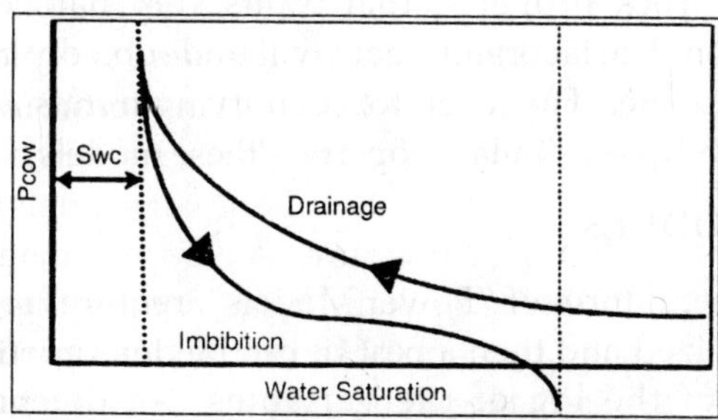

Fig. 2.8

Hysteresis effects have been demonstrated for relative permeability relationships. Generally, the imbibition curves are sufficient for oil-water systems.

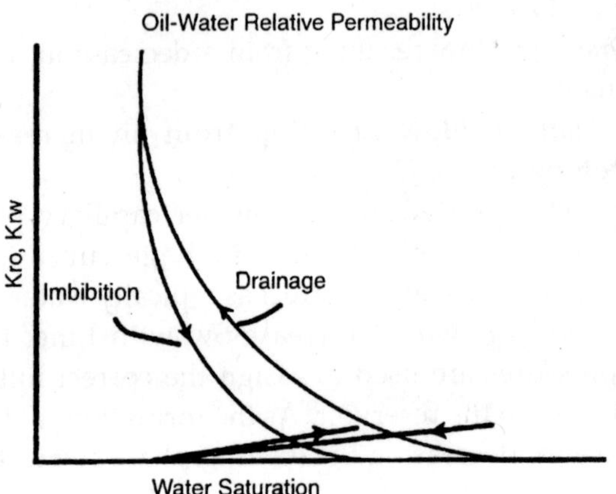

Fig. 2.9

Oil-gas relative permeability curves, as used in a simulator, can include endpoints Sgc, the minimum critical gas saturation for the start of gas flow, and Sgr, the trapped gas saturation resulting from gas being displaced by oil. The normal means to calculate this is to adjust the gas saturation, internally in the simulation programme.

EMPIRICAL FORMS OF RELATIVE PERMEABILITY

The following models have been proposed to attempt to generalize a rock property, that by its very nature, alludes generalization. The laboratory data will undoubtedly be the best source of estimates. However, for identifying trends, and better classifying groups of similar properties, these models have value.

POWER MODELS

The general form of "Power Models", requires saturations to be normalized and then a best fit can be determined. Linear regression of the log of the variables can determine the exponent, n.

If the domain of X is [a, b], then $X_N = (X-a)/(b-a)$

$$S_{O,N} = \frac{S_o - S_{or}}{1 - S_{wc} - S_{orw}}$$

$$S_{w,N} = \frac{S_w - S_{wc}}{1 - S_{wc} - S_{orw}}$$

$$S_{g,N} = \frac{S_g - S_{gc}}{1 - S_{org} - S_{wc} - S_{gc}}$$

$$k_{ro} = k_{ro,max} \cdot \left(S_{o,N}\right)^{no}$$

$$k_{rw} = k_{rw,max} \cdot \left(S_{w,N}\right)^{nw}$$

$$k_{rg} = k_{rg,max} \cdot \left(S_{g,N}\right)^{ng}$$

It is noted that the normalized saturations consider maximum S_w=1-S_{orw}

Other formulations consider maximum S_w=1-S_{gc} + S_{orw}

Chierici oil-gas exponential equations (as described in Referance 1),

$$k_{ro} = \exp\left(-A \cdot S_{g-N}^{L}\right)$$

$$k_{rg} = \exp\left(-B \cdot S_{g-N}^{-M}\right)$$

where the normalized saturations are defined as follow:

$$S_{g,N} = \frac{S_g - S_{gc}}{1 - S_{wc} - S_g}$$

and for oil/water relative permeabilites:

$$k_{ro} = k_{ro,max} \cdot \exp\left(-A \cdot S_{w,N}^{L}\right)$$

$$k_{rw} = k_{rw,max} \cdot \exp\left(-B \cdot S_{w,N}^{-M}\right)$$

where:

$$S_{w,N} = \frac{S_w - S_{wc}}{1 - S_{or} - S_w}$$

As stated, other models have been proposed. This area is closely tied to three phase relative permeability which I hope to ad in the near future.

GAS CONDENSATE RESERVOIR ANALYSIS

PHASE BEHAVOIR

Gas condensate field produce mostly gas, with some liquid dropout, frequently occurring in the separator. The phase diagram shows the retrograde gas field must have a temperature higher than the critical point temperature.

The vertical line on the phase diagram shows the phase changes in the reservoir, while the curve line shows these changes as the fluid cools going up the wellbore and into the separator.In both cases, liquids drop out as the pressure drops below dew point pressure.

With the retrograde condensate, the%liquid begins to increase to point "A" then decreases with further pressure declines. Thus the name "retrograde" meaning to retreat or go back. So first condensation and then vaporization occurs, and this vaporization can help in further recovery of liquids.

Any hydrocarbon above the dew point line is 100% gas. Any hydrocarbon above the bubble point line is 100% liquid. Hydrocarbons above the bubble point line and close to the critical point are volatile oils.

The *cricondentherm* is the maximum temperature which two phase flow can exist (maximum temperature on the dew point line of the phase diagram).

But a field may have both a oil leg and gas cap, which during depletion produces some condensate from the gas. Fields with active water drive may experience little pressure declines, so condensation occurs only at surface and a constant GLR would be expected.

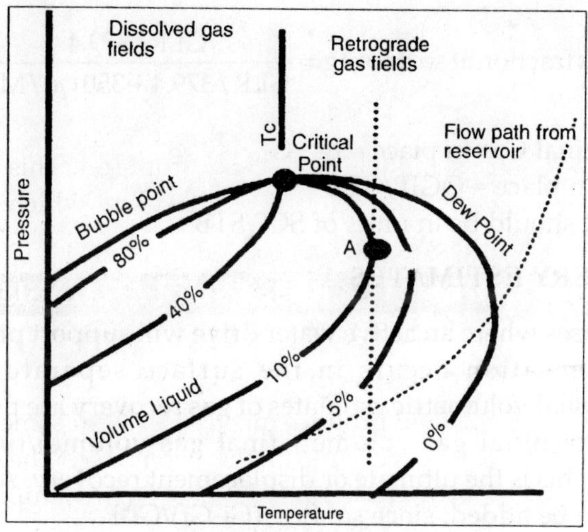

Fig. 2.10

SPECIFIC GRAVITY AND HYDROCARBONS IN PLACE

GLR = SCF gas to bbl of condenstate
GLR = SCF gas to STB condenstate
Specific gravity of produced well fluid is:

$$\gamma w = \frac{GLR \cdot \gamma g + 4585 \cdot \gamma o}{GLR + 132800 \cdot (\gamma o / Mo)}$$

$$\gamma o = \frac{141.5}{API - 5.9}$$

$$Mo = \frac{6084}{API - 55.9}$$

Cragoe's empirical relationship, use if Mo is unavailable.

G = 43560Aϕh(1-Swc)/Bg
Can use 43.560 and calculate MSCF to avoid large numbers.
Can also calculate G/Ac-ft.

$$G/(A \cdot h) = \frac{43560\phi(1 - Swc)}{Bg}$$

$$\text{Gas fraction at surface, fg} = \frac{GLR/379.4}{GLR/379.4 + 350 \cdot \gamma o/Mo}$$

Original Gas in place = fg · G
Oil in place = OGIP/GLR
GLR should be in units of SCF/STB

RECOVERY ESTIMATES

In cases where an active water drive will support pressure, so condensation occurs in the surface separators, the conventional volumetric estimates of gas recovery are possible, where Er=(nitial gas volume - final gas volume)/final gas volume. This is the ultimate or displacement recovery. A sweep factor can be added, since so: Er = (Gi-Gf)/Gi*F.

EMPIRICAL METHOD OF OIL RECOVERY USING THE API EQUATION

The API or "Arps" equation for oil recovery of solution gas drive reservoirs, is the result of an empirical multiple regression of 80 solution gas reservoirs. It is as follows:

$$Er = 41.815 \cdot \left[\frac{\phi(1-S_{wi})}{B_{ob}}\right]^{A} \cdot \left[\frac{k * 1000}{\mu_{ob}}\right]^{B} \cdot (Swi)^{C} \cdot \left(\frac{P_b}{P_a}\right)^{D}$$

Where:

- k = absolute permeability (md), note the original equation used Darcies
- μob = viscosity at bubble point, cp
- Bob = Bubble point pressure, RB/STB
- Pb = Bubble point pressure, psig
- Pa = Abandonment pressure, psig
- Er =% recovery from bubble point to abandonment
- A = 0.611, B=0.0979, C= 0.3722, D = 0.1741

More recently, the Arps equations were presented in Basic Applied Reservoir. A basic assumption in multiple regression is that the variables are independent, which means that the value of porosity should not affect initial water saturation or permeability.

Of course these are related variables. Viscosity, Bop and bubble point pressure are also related variables. This introduces a systemic error in the analysis. If necessary to use, it would best to programme in a spreadsheet, remembering that a/b^c is not the same as (a/b)^c. If using a calculator, calculate the four products first and then multiply to keep your work organized.

WATER DRIVE RESERVOIRS

The API equation for waterdrive reservoirs is similar in form, based on multiple regression:

$$Er = 54.898\left[\frac{\phi(1-S_{wi})}{B_{ob}}\right]^{A}\left(\frac{1000\cdot k\mu_{wi}}{\mu oi}\right)^{B}(S_{wi})^{C}\left(\frac{p_b}{p_a}\right)^{D}$$

where A= 0.0422, B= 0.77, C= -0.1903 and D= -0.2159.Similar comments as before apply to the validity of the regression analysis.

ADDITIONAL REGRESSION BASED EQUATIONS

- Jacoby Correlation for gas condensate recovery.

$$N_F = -0.061743 + \frac{143.55}{R_{pi}} + 0.0001218T + 0.0010114(°API)$$

$$G_F = -2229.4 + 148.43\cdot\left(\frac{R_{pi}}{100}\right)^{-0.2} + \frac{124130}{T} + 21.831(°API) + 0.26356\cdot p_o$$

Where N_F = Cumulative liquid Production, STB/(RB of hydrocarbons pore volume) and G_F = Cumulative gas production, scf/RB of hydrocarbons pore volume) T is in °R and R_{pi} = initial producing gas/condenstate reatio, scf/STB

MATERIAL BALANCE EQUATIONS

$$N = \frac{\text{Re servoir Pr oduction, injection and water inf lux}}{\text{Pr essure dependent terms related to fluid exp ansion (and possible pore volume compaction}}$$

BUBBLE POINT

$$C_e = \left[\frac{c_w S_{wi} + S_o C_o + C_1}{1 - S_{wi}}\right]$$

$$N = \frac{N_p}{C_e \Delta p}\left(\frac{B_o}{B_{oi}}\right)$$

Or :

$$N = \frac{N_p (B_o)}{B_1 - B_{oi} + B_{oi} c_e \cdot \Delta p}$$

For water production, water injection and water influx, change numerator to:

$$N_p B_o + \left(W_p - W_i\right) B_w + W_e$$

Note that water influx is in cumulative reservoir barrels. Excluding compressibilities will result in overestimation of OOIP, particularly in friable sands with high formation rock compressibilities. We will address means of calculating the water influx in Part II of this discussion.

BUBBLE POINT

As reservoir pressure declines below bubble point pressure, free gas starts to evolve and a free gas phase develops. A secondary gas cap may or may not form. The movement of free gas will depend on several factors, one is the gas-oil relative permeability curve. A critical saturation may need to develop before the onset of gas movement. The free gas may remain distributed throughout the reservoir as insoluble bubbles. The following material balance equations apply in both cases:

$$N = \frac{N_p \left(B_o - R_s B_g\right) + G_F B_g}{B_o - B_{oi} + B_g (R_{si} - R_s) + m B_{oi} \left(\frac{B_g - B_{gi}}{B_{gi}}\right)}$$

Or :

$$N = \frac{N_F\left[B_t + B_g\left(R_F - R_{si}\right)\right]}{B_t - B_{ti} + mB_{oi}\left(\frac{B_g}{B_{gi}} - 1\right)}$$

If water is produced, water is injected or water influx occurs, the numerator can be changed to:

$$N_F\left[B_t + B_g\left(R_f - R_{si}\right)\right] - W_e + B_w W_F$$

Where:

- N = Intial oil in place(STB)
- Np = Cumulative oil (STB)
- B_{oi} = Initial Oil FVF (Robbl/STB)
- B_o = Oil FVF at pressure P_1 at time T_1(Rbbl/STB)
- B_g = Gas FVF at time T_1 (Rbbl/STF)
- $B_t = B_o + B_g(R_{si} - R_s)$
- R_F = Cumulative produced GOR at time T_1
- $R_F = F_G/N_F$
- G_F = Cumulative gas produced by time T_1
- m = Pore volume of gas cap/pore volume of oil zone, 0-dimensionless

Notes: a) Some sources present 3 MBE, one for above bubble point, one for saturated conditions without gas cap and one with gas cap. I did not think this was necessary since m = 0 without a gas cap. b) To minimize constants, I have used a Bg calculated as Rbbl/SCF instead of Rcf/STB, so divide the unitless Bg by 5.615, or multiply by 0.1781 c) I note in the 1982

GAS MATERIAL BALANCES AND STRAIGHT LINE METHODS

GAS-IN-PLACE WITH WATER INFLUX

In prior discussion of MBE, we multipled Gp by B_g, the formation volume factor to obtain a reservoir volume in either cubic ft or bbls. Remember the constant 0.02829 was used to calculate Bg in cubic ft, and 0.00504 for barrels.

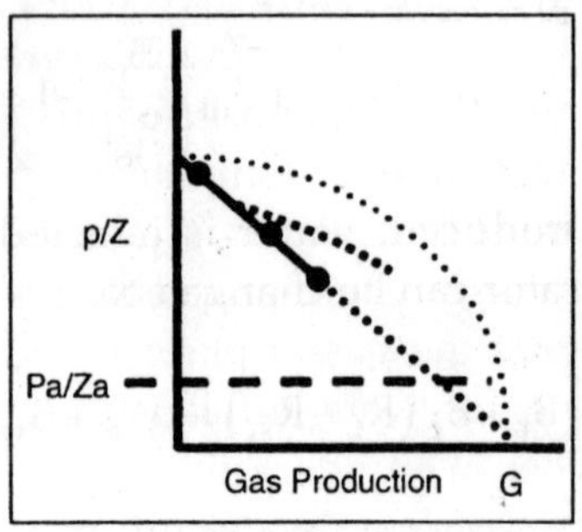

Fig. 2.11

Recovery of a dry gas with adequate production history and no water influx, can be determined by the P/Z verses Gp plot. Overestimation of OGIP and ultimate recovery is possible in the case of a geopressured formation or water influx. The pressure support from an aquifer may not be apparent at first. Assuming no water influx when one is present can result in over estimation of both gas in place and ultimate recovery.

From straight line MB methods, we intend to calculate GIIP and cumulative water influx, We. This is a function of aquifer's geometry and transmissibility T = Kh. Aquifers may initially provide complete support or partial support depending on these characteristics.

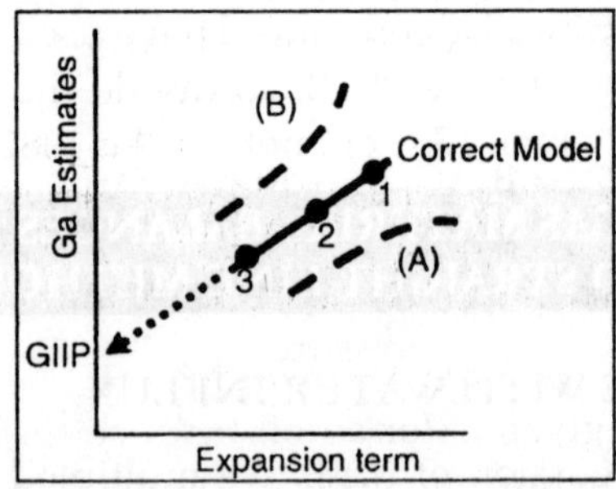

Fig. 2.12

The next graph has purposely been left unlabeled. This of course, is the straight line approach to material balance, and the correct model has the right cumulative water influx, We. The points are labeled 1, 2 and 3, to show the time sequence. You can think of water influx as the opposite of production.With Line (B), it appears that we will over estimate gas in place.

Therefore, we must be using an We that is too high or too low? Think for a moment before proceeding. Congratulations! We is too low. Just as the gas-in-place is too high when we neglect completely neglect water influx in the P/Z method, GIIP is overestimated when we underestimate it. The straight line equation for gas reservoir with water influx is:

$$\frac{G_p B_g + W_p B_w}{\left(B_g - B_{gi}\right)} = \frac{C \cdot f(p,t)}{B_g - B_{gi}} + G$$

in the Y = mX+b format, where the line has slope C and intercept G. The *f(p,t)* term is the aquifer model, which can be either steady-state or unsteady state.

STEADY-STATE MODEL FOR AQUIFER

The steady-state model means that the aquifer keeps up with production. Perhaps initially, the field declines in pressure and rates drop off. At this point, the total voidage in the reservoir equals the influx from the aquifer and the pressure stabilizes. In the steady state model, water influx rate is in proportion to the pressure drop. Under this condition, the water influx rate must equal the gas and water rate, all expressed in reservoir units.

$$\Delta W_e/\Delta t = B_g \Delta G_F/\Delta t + B_w \Delta W_F/\Delta t \text{ and}$$

$C = \Delta W_e/(p_i - p_s)$ where p_i = initial pressure and p_s = stabilized pressure

The example given is for an oil field, so the released free gas and oil production must be accounted included in the calculation.

UNSTEADY-STATE AQUIFER

It is based on the same equation as the well test equation, the radial diffusivity equation. Similar to the aquifer model are the diffusivity constant, which divides storage terms (porosity, total compressibility) by transmissibility (permeability/ viscosity). The HE model was later extended by Coats to include bottom water drive. Vertical permeability was added to the model.

OTHER MODELS

Many times, given a structure map, aquifers are part of the reservoir model using larger than normal grid cells. The same problem of identifying properties, particularly aquifer extent and permeability persists.

GEOPRESSURED RESERVOIR

Will add a link to this topic soon. Note that the geopressured reservoir appears at first to be an enormous field by conventional P/Z techniques.

FINAL COMMENTS

Theoretical models of aquifers are the same for gas and oil bearing reservoirs, however oil reservoirs add complications due to gas evolution from oil. The straight line methodology attempts to identify the GIIP and C (aquifer constant) that would explain the pressure decline in the field. The alternative is history matching with reservoir simulation. With increases in the speed of these programmes, the use of MB straight line methods are in decline, at least in my experience.

The modern approach to simulation is to model flow up the wellbore and provide a complete forecast including well head pressures. Avoid material balance methods when tank like assumptions will not hold. However, the MB method is far from dead. The simplified model may help assist or give further support to a history match of a gas-aquifer system. Interestingly, material balance methodology is the preferred method for coalbed methane analysis for gas in place calculation.

UNSTEADY STATE AQUIFER REPRESENTATIONS

INTRODUCTION

Analyses of well test and aquifer pressure responses share a common theoretical basis, the diffusivity equation. Both seek to identify potentially non-unique model parameters. Unlike well testing, the aquifer analyses must work from observational data, very different from a designed test.

Also, the time span is commonly in terms of several years. The representation of the aquifer may be used in calculation of original gas or oil in place, or in the forecasting of production.

The model requires assumptions or estimates of the following:

- Geometry- radial, linear, bottom water drive
- Perneability and viscosity (Transmissibility)
- Porosity and compressibility (Storativity)

The irregular shape may be accomodated by either numerical simulation or by using an apparent radius and angle which provide a reasonable equivalent shape. This is discussed in Craft and Hawkins. Similarly, for non-uniform properties, calculation of aquifer volume, and adjusting porosity may be one method. Simulation is another alternative.

We will review three Van Everdingen-Hurst unsteady state models, which are:

1. VEH Infinite radial model
2. VEH finite radial model
3. VEH Linear model

There are other forms including the bottom water drive, which we hope to add later. Additionally, the Carter-Tracy model is used in reservoir simulation. Fetkovich method is an approximation to VEH, with the assumption of pseudo steady-state and was easier to hand calculate. However, today when material balance is done on the computer, the need for simplified solutions is less.

CALCULATION OF CUMULATIVE WATER INFLUX

The starting point is really the conservation of mass equation and Darcy's equation, which result in the diffusivity equation as presented, and made dimensionless by pressure, time and radius linear transforms..

$$\frac{\partial^2 p}{\partial r^2}+\frac{1}{r}\frac{\partial p}{\partial r}=\frac{\phi\mu c_t}{0.0002637\cdot k}\frac{\partial p}{\partial t} \text{ or : } \qquad \frac{\partial^2 p_D}{\partial r_D^2}+\frac{1}{r_D}\frac{\partial p_D}{\partial r_D}=\frac{\partial p_D}{\partial t_D}$$

where time is in hours. Use 0.00633 as constant if time is in days

Dimensionless time, pressure and radius:

$$t_D=\frac{0.00633\cdot kt}{\theta\mu c_t r_r^2} \text{ with t in terms of days.}$$

For constant rate conditions at reservoir/aquifier boundary:

$$P_D=\frac{0.00708\cdot kh(p_i-p)}{q\mu}$$

For Constant terminal pressure conditions,

$$P_D=\frac{p_i-p}{p_i-p_r}$$

Dimensionless radius is:

$$r_D=r/r_\gamma$$

- Infinite aquifier (suitable in general for r_o/r_g >10

$$q_D=\frac{q_w\mu}{0.00708\cdot kh\Delta p}\text{, where } q_D \text{ and } q_w \text{ are inf lux rate}$$

$$\text{Defining } W_D=\int_0^{t_D} q_D dt_D=\frac{W_e}{1.119\phi c_t h r_r^2 \Delta p},$$

Note: constant 1.119 = 0.00708/0.00633.

$$W_e=1.119\phi c_r h r_r^2 \Delta p W_D$$

$B=1.119\phi c_r h r_r^2$ or if the angle subtended by the reservoir $\leq 360°$

$$B=1.119\phi c_r h r_r^2\left(\frac{\theta}{360}\right) \text{ where B is in units RB / psi}$$

$$W_e(t_n) = B\sum_{i-1}^{n} \Delta p_i W_D(\Delta t_D) \text{ which result in a series of summations}$$

$$\text{where } \Delta t_D = \frac{0.00333k\Delta t}{\phi\mu c_t r_r^r} \text{ where t is in days, radius in ft.}$$

$$\Delta p_1 = 1/2(p_i - p_1)$$

$$\Delta p_2 = 1/2(p_i - p_2)$$

$$\Delta p_3 = 1/2(p_1 - p_3)$$

Or:

$$\Delta p_i = 1/2(p_{i-2} - p_i) \text{ and}$$

$$p_{-1} = p_0 = p_i$$

- For limited aquifiers:

$$W_e(t_n) = B\sum_{i-1}^{n} \Delta p_i W_D(\Delta t_D, r_D)$$

Requires tables or equations. Dimensionless radius = r_o/r_γ. Equations for t_D and B are unchanged from infinite aquifer case.

- For linear aquifers
  L= Distance from reservoir/aquifer interface (t is in days)

TRADITIONAL DECLINE CURVE ANALYSIS

Exponential, Harmonic or Hyperbolic Decline

The days of plotting rates on semi-log graph paper are long gone. In the industry, commercial software programmes are used with extensive database capabilities to quickly develop forecasts. In general, the exponential decline is the most commonly used method. Perhaps in the beginning, it was as matter of convenience.

Today, with everyone using software generated decline curves, all methods are equally convenient. The graph below shows all three decline equations fit nearly exactly to the first 2 years, and produce noticeably different forecasts. If all three decline equations match the historical data, then the exponential decline forecast will show the most decline in rates, hence provide the most conservative forecast.

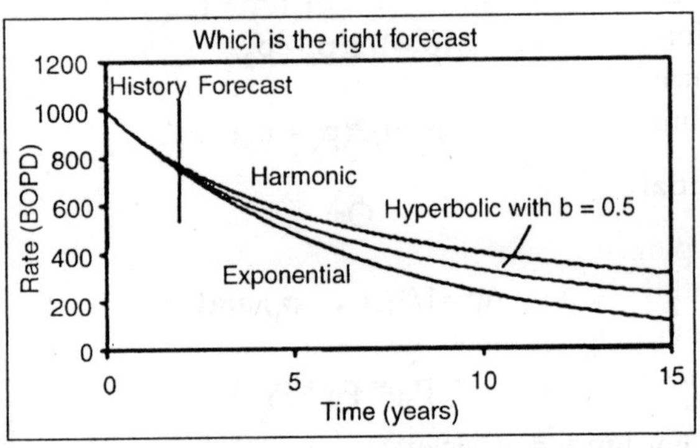

Fig. 2.13

Conservative forecasts are probably most appropriate in the early stages of production where there are more unquantifiable factors that tend to reduce the field rate rather than increase them in the future. This includes down time related to facility and well maintenance. Secondly, if the decline curve analysis is to be used for reserves calculations, the forecast needs reflect a "reasonable certainty" standard.

The hyperbolic curve requires estimation of both Di and the exponent "b." Hyperbolic decline curves have been used in many cases, particularly in the late history of pressure supported production. Finally, the cumulative recovery should result in a reasonable recovery factor, based on volumetrics.

Most computer programmes allow multiple forecasts to be made, and sensitivity analysis on the computed variables. If the objective of the forecast is an upside projection, useful in facility design, then harmonic and hyperbolic methods may be employed.Also, note that the rate, BOPD, can be MSCF gas. In fields under pump and producing high volumes of water, then a linear relationship between log of water cut and Np would reflect harmonic decline. Linear relationship between water cut and Np would suggest a exponential decline.

Notation

qi = Initial rate, q(t) = rate at time t, Q(t) = Cumulative Production, Di = decline rate, b = used in hyperbolic decline, b = 1 for harmonic. Keep units consistent. Decline rate is not annual decline, but used as a parameter in the equation.

Exponential

The plots on the left side are based on a common 15 year rate for all three methods. DCA is applicable when the plot of rate vs. log(time) has a constant slope, Di. Plot of rate vs. cum oil, also has a straight line. The exponential model is:

$$-(dq/dt)\cdot(1/q) = Di$$

$$q(t) = qi\cdot\exp(-Dit)$$

$$Q(t) = (qi - q\,(t))/Di$$

Note that when Di is obtained from a semi-log plot, it must be divided by 2.3 to convert to natural log scale.

Harmonic

Harmonic decline is a special case of the hyperbolic curve for b = 1. The slope of ln(q) vs cumulative production plot is equal to D_i/q_i or on log base 10, the slope is: $D_i/(2.3\cdot q_i)$

$$q(t) = q_i/(1+D_i t)$$

$$Q(t) = \frac{q_i}{D_i}\ln(1+D_i t)$$

$$\ln[q(t)] = \ln(q_i) - \left(\frac{D_i}{q_i}\right)Q(t) \text{ or}$$

$$\log_{10}[q(t)] = \log_{10}(q_i) - \left(\frac{D}{2.3 \cdot q_i}\right) \cdot Q(t)$$

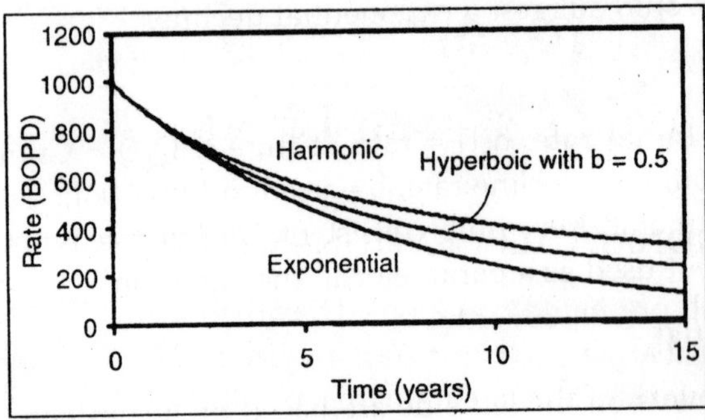

Fig. 2.14

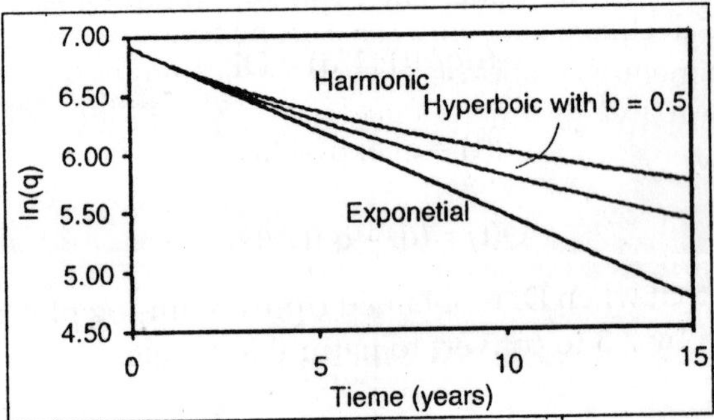

Fig. 2.15

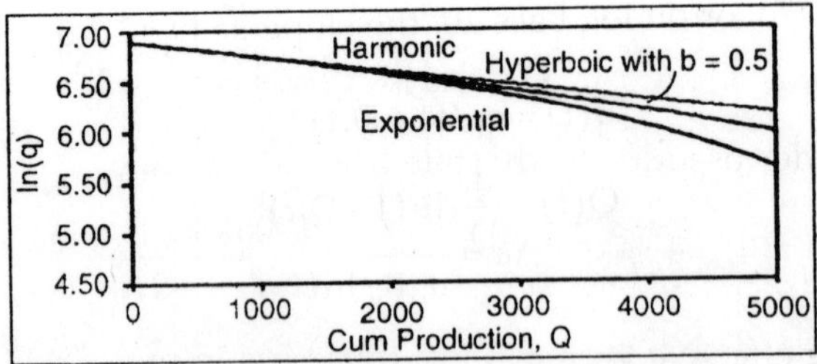

Fig. 2.16

Hyperbolic

Hyperbolic decline uses two parameters, Di and b.

$$q(t) = \frac{q_i}{(1+bD_i t)^{1/b}}$$

$$Q(t) = \frac{q_i^b}{D_i(b-1)}\left[q(t)^\alpha - q_i^\alpha\right] \text{where } \alpha = 1-b$$

ADDITIONAL CALCULATIONS WITH EXPONENTIAL

In all presented equations, D_i is the normal decline. Some authors use "a" for normal declines. Normal decline is continuous over timeor cumulative production. Effective decline is the fractional rate decline from time t_1 to time t_2, or simply $d_{eff} = (q_1 - q_2)/q_1$ and is easily calculated from the above equations. For exponential decline, $d_{eff} = 1 - \exp(-D_i)$. A constant effective decline exists only for exponential declines. Given an abandonment rate and decline rate, the time to abandonment and cumulative production at abandonment can be calculated, as follow,

$$t_a = 1n(q_i/q_a)/D_i$$

and

$$Np = (q_i - q_a)/D_i$$

Well Inflow Analyses

The well's productivity index (PI or *J* is common) is:

$$PI = q/\left(\overline{P}_R - P_{wf}\right)$$

Under pseudo-steady state *(pss)*, a calculated PI is:

$$PI_c = J_c = q_o/\Delta p = \frac{0.00708k \cdot k_{ro} h}{\mu_o B_o \ln\left(r_e/r_w - 3/4 + s\right)}$$

The effective permeability-thickness in md-ft ($k \cdot k_{ro} h$) and skin (*s*) are estimated based on well test analyses. The external radius, r_e, is the radius in feet to the no-flow boundary, based on the approximate drainage area of the well. The radial flow equation assumes a bounded circular drainage area, a fully penetrating well, the well is vertical and flow is single phase. This is in addition to the assumed conditions in Darcy's law. A zero skin damage PI (or ideal PI) would be:

$$PI_i = J_i = q_o / \Delta p = \frac{0.00708 k \cdot k_{ro} h}{\mu_o B_o \ln(r_e / r_w) - 3/4}$$

Presentation of pressure transient test results for developed fields should include the history of PI's. A time plot showing PI's and idealized PI's (or rates and ideal rates) for a particular well and surrounding area, should be part of reservoir monitoring to identify common problems.

A positive skin factor indicates there is a pressure decline in the near vicinity of well that is more than expected based on the radial flow equation. A positive skin factor does not necessarily indicated formation damage. However, as a conceptual model, if we consider the reservoir has been damaged out to a radius, r_s, then the skin of this zone can be calculated based on the reduced permeability:

$$s = \left(\frac{k}{k_s} - 1 \right) \ln \left(\frac{r_s}{r_w} \right)$$

The difficulty in the above equation is knowing the damage radius. From the calculation of skin, the additional pressure drop due to skin is:

$$\Delta p_s = 0.87 \cdot m \cdot s$$

m=162.6·qBμl (kh) slope of the nfinite-acting period on the horner plot.

The specific PI per foot of producing interval, or: J_s = J/h = bpd/psi/ft allows comparing wells with different producing intervals. Also, the steady-state radial flow equation has been used by some authors:

$$PI_i = J_i = q_o / \Delta p = \frac{0.00708k \cdot k_{ro}h}{m_o B_o \ln(r_e / r_w)}$$

While technically incorrect, the effect of disregarding the -3/4 may be insignificant. The PI as calculated with the pseudo-steady equation or with the steady state equation, may consider multi-phase flow, with each flow converted into reservoir barrels by its respective rate. The multiphase flow equation substitutes for k_o/μ and $q_o \cdot B_o$, the total transmissibilities and reservoir withdrawals as:

$$(k/\mu)_t = (k/\mu)_o + (k/\mu)_w + (k/\mu)_g$$

$$q_t B_t = q_o B_o + (1000 \cdot q_g - R_s \cdot q_o) \cdot B_g / 5.615 + q_w B_w$$

The constant of 5.615 is used to convert cubic feet to barrels. Note that the PI as determined from production tests may also be calculated based on total volumes of produced water, oil and gas, all stated in terms of stock tank barrels.

Flow Efficiency

Flow efficiency is the actual PI/(theoretical PI without skin). In terms of drawdown, it is:

$$FE = \frac{\bar{P}_R - P_{wf} - \Delta P_{skin}}{\bar{P}_R - P_{wf}}$$

Flow efficiency for two flow tests will be included later in this discussion.

Reservoir Boundaries

Pseudo- steady pressure declines in irregularly shaped reservoirs have been determined using superposition of the transient flow equation:

$$PI = J_i = q_o / \Delta p = \frac{0.00708k \cdot k_{ro}h}{\mu_o B_o \left[0.5 \cdot \ln\left(\frac{2.2458}{C_A} \right) + 0.5 \cdot \ln\left(\frac{A}{r_w^2} \right) \right] + s}$$

A is given in square feet (43560 sq ft = 1 acre). The geometry constant, 0.5·ln

$$\left(\frac{2.2458}{C_A}\right)$$

THE PSEUDO-STEADY STATE PERIOD

It is convenient to subdivide these flow periods (or shut-in periods) into 3 regions: early time, middle time and late time. Early time is affect by well bore storage.

The middle time is the semi-log straight line pressure response on the drawdown test. Late time is affected by well boundaries, and this is when the pressure response transitions from infinite acting to pseudo-steady states. The time to the start of pseudo-steady state is obtained from the shape factor, under the column, "exact for t_{DA}" which can be converted into time in hours by:

$$t_{DA} = \frac{0.000264kt}{\phi\mu c_t A}.$$

The start of pseudo-steady state is when the farthest boundary has been felt at the wellbore. A reservoir that has transitioned to pseudo-steady state will have a linear decline in pressure with constant production.

The approximate and the exact times to reach pseudo-state, from the shape factor chart, for a circle are fairly close (0.06 and 0.10). However, in cases where the well is positioned much closer to one side, the time from "approximate beginning" and "exact beginning" is much longer.

This can also be seen in the plots of dimensionless pressure verses time. The pseudo-steady state period can, at least in theory, provide estimates of the reservoir pore volume:

$$m = \frac{-0.234qB}{V_F c_t}$$

EMPIRICAL WELL INFLOW EQUATIONS

Vogel's Inflow Curve

Well inflow equations do not account for the phase change in solution gas reservoirs. In a solution gas drive, there is expansion of the hydrocarbons below bubble point, which is beneficial because it adds energy to the system. Gas liberation in a solution-gas drive is also detrimental to oil production because it lowers the effective permeability of oil.

Vogel's curve is entirely empirical, based on a series of single well solution-gas simulation runs, with varying oil properties

$$\frac{q_o}{q_{o,max}} = 1 - 0.2\left(\frac{P_{wf}}{\overline{\overline{P}}_R}\right) - 0.8\left(\frac{P_{wf}}{\overline{\overline{P}}_R}\right)^2$$

The Vogel curve, will provide the maximum rate possible given the test's rate, reservoir pressure and flowing bottomhole pressure. Future additions—Partial completions, gas well inflow relationships, isochronal tests.

WATERFLOOD PREDICTION AND MONITORING

Prior to the waterflooding or pressure maintenance by water injection, the most common means of prediction is reservoir simulation. In fact, I have never known a waterflood project to be considered without a simulation model. I apologize at the onset, to all those die-hard classic approach followers, but I really think that recovery by Stiles or Dykstra-Parson's method is a matter of history.

As for streamline simulation, it is the tool, which for many years, got no respect, but continued to be improved and generate accurate forecasts. It is easier to history match through automated methods. The savvy engineer can pick the tool which will best accomplish his goals.

Classic Approach- for What it's Worth

The classical approach of calculating a "ballpark" recovery

factor can be done, by first calculating the permeability variation, Dykstra Parson's $V=(k_{50} - k_{84})/k_{50}$ as taken off the log-normal plot. Alternatively, the k84 value can be calculated from one standard deviation less than the median of the log(k) values.

Recovery also requires the mobility ratio, $(k_{rg}/\mu_g)/(k_{ro}/\mu_o)$ for gas displacing oil, or $(k_{rg}/\mu_g)/(k_{ro}/\mu_o)$ for water displacing oil. Gas viscosity is at least an order of magnitude higher than oil, mobility is greater than 1 and gas will finger through the oil phase. This is an example of an unfavourable mobility ratio. Water displacing a light oil result in a low mobility ratio, or favourable displacement.

A high mobility means a high velocity, so high velocity/ low velocity results in fingering and poor recovery. Also remember ING:/ED, displacing/displaced phase. Or if you must, remember fat Ingrid on top of Ed at least for gas displacement. Ok. I can estimate waterflood recovery using M, V, Sw, and WOR.

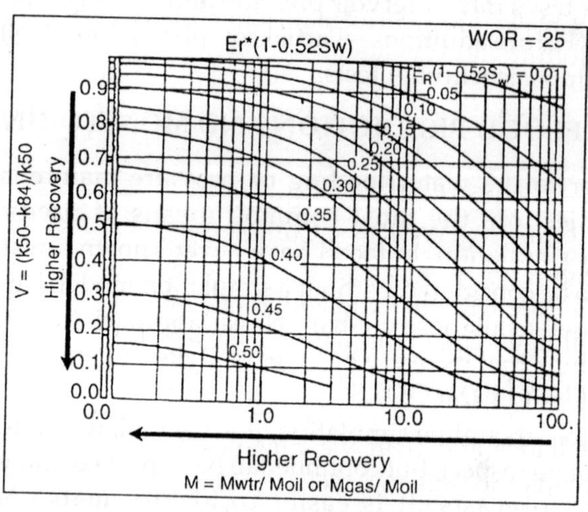

Fig. 2.17

So, let say we have an oil with viscosity of 0.5 cp (pretty average), we'll assume that the ratio of krw/kro endpoints is 0.3, and water viscosity = 1, so M is 0.3/0.5 or 0.6. If V = 0.8, the value off the graph is 0.25. A value of Sw = 10 to 30% is only going to slightly increase so a 22- 24% recovery factor is calculated for waterflooding.

MONITORING/FORECASTING WATERFLOOD PERFORMANCE

Remember I had nothing good to say about use of "classical methods" for waterflood predictions prior to implementation as I much preferred simulation models. However, post implementation is a different story. All waterfloods will deviate from simulated results. Bubble maps of injectors are excellent visualization tools.

Given thickness, moveable oil saturation (1-Swc-Sorw) and porosity, the ideal swept zone from the injector can be identified. In a depleted field, there is a period of fill-up or resaturation of the gas phase. Careful adjustments to the injectors can optimize this process. Ideally, injected reservoir volumes of water should equal the total voidage by producing wells. Plots useful for forecasting waterflood performance include the X plot

- Plotting *Er vs X,* where

$$X = \ln\left(\frac{1}{f_w} - 1\right)\frac{1}{f_w}$$

- Plotting log(WOR) vs. Cumulative Recovery "cum-cut plot"
- Oil rate with an exponential decline with time
- Oil rate verses cumulative recovery.

SOPHISTICATED TOOLS

Surveillance and field visualization tools help engineers identify trends and potential wells needing workovers. Software packages display an abundance of information related to the well history for better decision making. Certainly, the anticipated production from streamtube, simulation or decline curves can be compared with actual production, to cull out wells needing attention.

Fence diagrams coupled with geological interpretations are invaluable in understanding waterflood behavoirs. More practical information can be found in SPE Monogram 11, entitled The Design Engineering Aspects of Waterflooding, by Rose, Buchwalter and Woodhall.

EVALUATION OF EXPLORATION PROSPECTS

INVESTMENT ANALYSIS

Exploration prospect evaluations differ from many other investment analyses, because they are entirely dependent on an uncertain event- a hydrocarbon discovery that meets the threshold of being commercial. Prior to drilling, due to this uncertain event, any analysis is inherently speculative. Sometimes seismic data can suggest the presence of hydrocarbons, however true discovery occurs only by drilling wells and running tests.

The common benchmarks for evaluation of any project are net present value (NPV), discounted cash flow rate of return (DCF-ROI), present value index (PVI), and payout. Payout can be based on discounted or undiscounted cash flows. In theory, investments should be made in accordance to the "cost-benefit" ratio to maximize asset returns.

Thus acquisitions should be made with the projects with the highest present value index (NPV/Investments). This simple concept is difficult to apply when a project proceeds in a series of stages, and each stage is contingent on success of the previous stage. The means of investing in exploration areas are numerous.

They include acquisitions, farm-ins and competitive bidding for exploration acreage. In all cases, there are large upfront costs and uncertainty. Companies can limit their exposure to the large initial exploration costs by forming operator groups composed of a number of partners.

NEW DRIVERS FOR CHANGE

Investment analysis of oil/gas ventures exists in a microcosm of an industry experiencing an incredible transformation. It was recognized in the 1990's that new drivers would play an increasing role in the future exploration efforts.

The "Drivers of Change" enumerated have all come to fruition:

- Volatility of Oil Prices
- Deintegration of upstream resources and downstream sources
- New technology – principally 3-D seismic, horizontal drilling, and tension leg platforms.
- Communication revolution
- Political transformation and breakdown of national frontiers
- Growing global environment awareness
- Ownership and governance –Expanded state sponsored NOC investments
- Deregulation

The most powerful drivers for the 21st Century are growing global environment awareness and new technology. This awareness that companies needed to examine every opportunity in the world and the new technology for deep water exploration translated into a future of need for large well funded and international oil companies. This lead to nearly a decade of mega-mergers.

BP/Amoco 1998	Conoco/Philips 2002
Exxon/Mobil 1999	Devon/Ocean Energy 2003
Chevron/Texaco 2001	

Consolidation of the seismic processing, drilling and other oil service companies has been equally rapid, with a partial list provided:

Santa Fe/Global Marine 2003	National Oilwell/Varco 2005
Baker Hughes/Western Atlas	Baker Hughes/B.J. Services – Announced Aug 2009 in progress
Santa Fe/Global Marine 2003	
Western Geco/Schlumberger 2006	

The impacts on this structural transformation meant:

- More competition for the high value exploration blocks around the globe with NOC
- Major oil companies interested in operating the block, rather than being participant, and
- A few mega-service companies partnering with the oil companies at every phase of exploration and production in every location where E&P exists.

EVALUATION OF EXPLORATORY PROSPECTS

Exploratory prospect evaluation requires risk evaluation. There is no single method for analysis.

The probability of the existence of hydrocarbons in a play depends on:

- Confinement or closure of the deposit
- Adequate deposition of organic material to form oil
- Right conditions (time, pressure and temperature) to become oil or gas

If each factor is assigned a probability, then P{hydrocarbons exists} = $p_1 \cdot p_2 \cdot p_3$ Often this is referred to as the "geological risk." This does not describe the chance of a commercial discovery.

For a commercial discovery, the discovered accumulation must meet certain threshold criteria:

- Sufficient quantity of recoverable hydrocarbons (HC)
- Reservoir productivity is sufficient to produce hydrocarbons at commercial (economic) rates
- Hydrocarbon discovered is commercially marketable-this may rule out a large gas, condensate, or volatile oil discovery as commercial.

Also, heavy oils discovered offshore may be considered non-commercial. P{field is commercial given HC present} = $p_1 \cdot p_2 \cdot p_3$ Some economic analyses will lump these factors into a "commerciality risk factor" however if possible, it is worthwhile to identify each one separately. P{Commercial discovery} P{Commercial} · P{hydocarbons exists} Now, consider that the chance of success of each of these six factors is 95% or a 1 in 20 chance of being absent. The chance of hydrocarbons being present is 86%, but the chance of a commercial discovery is 74%. If all factors are 50%, the chance of hydrocarbons drops to 12.5% or 1:8, and the chance of a commercial discovery is 1.5%.

If in this example, we can eliminate the chance that the hydrocarbon is marketable, the chance of a commercial discovery now is 3% or 1:33. Now, an exploration programme of a block will typically involve more than one well, so the chances of one or more successes will in theory, increase as exploration programme continues. The probability of having one success after "n" wells are drilled, each with a success probability, P_s, can be calculated using the binomial distribution.

MONTE-CARLO SIMULATION TO DETERMINE RECOVERABLE OIL

The expected recoverable oil in an accumulation hydrocarbons as risked is:

$$\text{Exepected Recoverable oil} = P_s \cdot (OOIP) \cdot E_R$$

$$\text{where}\, OOIP = 7758 \cdot \frac{A \cdot \varnothing \cdot h}{B_o}(1 - Sw)$$

Classical risk analysis assigns probability distributions to the variables including E_R. Typically, the triangular distributions are used. Triangular distribution are simple to understand and provide a limited domain for the random variable. Other distributions such as the uniform and normal can be used. From Monte-Carlo simulation, a distribution of $OOIP \cdot E_R$ can be generated. Given P_G and recoverable oil, the expected reserves for the discovered hypothetical field within the block can be calculated.

ECONOMIC EVALUATION

At this point, it is necessary to construct various scenarios to identify an expected return on the prospect. If the contract establishes a mandatory exploratory programme, then this is given in our model as a 100% certainty cost, along with signature bonus. If no oil is found, then this establishes our "unsuccessful scenario" or maximum loss. The "success case" may involve numerous evaluations, considering various drilling costs and facility costs. Ultimately, each of these scenarios must be assigned a probability.

Under a Production Sharing Contract, cost recovery provisions allow an Operator to achieve a more rapid payout of the investment, however as the project becomes more profitable, the PSC mandates that a greater share of revenues go to the government. The economic evaluations of the various scenarios are very important as they identify the full range of economic gain and loss. The private oil company, may after examining the potential for disaster, not proceed unless it can reduce its working interest.

WEAKNESSES

The risk factor approach has a number of weaknesses, including:

- Estimation of probabilities and distributions are highly subjective and
- Each risk factor is considered to be independent and stationary (time invariant) contrary to the nature of the factors.

To improve the quality of estimation, companies often rely on outside geologic and seismic consultants, to provide additional opinions. The assumption of independence of variables is done to simplify analysis. The short comings of this form of analysis should be recognized early on. As a real world example, an exploratory block in Brazil had shown to have an extensive structure based on seismic.

As there was no prior wells, the seismic structural map was in time units. As the drilling progressed, the sands became increasingly more compacted with no signs of hydrocarbons. The top of target was continually prognosticated to be considerably deeper due to the higher velocities. Drilling costs rose due to compaction. The well's chance of success was quickly diminishing due to the interaction between risk factors. At times, the interdependence of risk factors is responsible for positive results. A high porosity will impact the quantity of oil found and improves the productivity of the well. Monte-Carlo simulation can be performed using correlated variables. This may be helpful in the economic simulation of prospects. A higher oil price forecast is correlated with higher drilling costs, and facility costs.

TIME INVARIANCE

The binomial distribution works when the success probability is a fixed value. There is a philosophical argument of whether, with more exploratory drilling, we will: a) become more successful, because every dry hole shows us our errors or b) become less successful, because we progressively explore the less prospective structures. Both perspectives have merit. However, with improved seismic techniques, the latter argument of progressively less prospective structures, I believe, holds more support.

OTHER APPROACHES

Decision tree diagrams can be used either as an alternative or to complement Monte-Carlo studies. It is a very effective tool in showing risks. It shows many ways to construct diagrams to describe economic analyses. Sensitivity analyses, including best/worse case scenarios and impact of variables, provides added insight to the analyses.

An approach of calculating "thresholds" in modeling economics also adds insight. If a project is uneconomic, under what conditions might we consider it to be economic. What possible changes can be made to the Production Sharing Contract (PSC) terms to make exploration."

Exploration decisions do not rely on a few simple metrics. To maximize assets, one should choose investments in accordance to the NPV/investment or PVI. In comparing two prospects, one with an expected PVI of 2 and a second one with a PVI of 1.6, an Operator opts for the lower PVI. Why? The investment of the higher PVI and the potential loss is much higher. The Operator is risk averse. The role of "risk acceptance/ aversion" is part of utility theory.

The expected economic returns do not tell the full story. An investment without sufficient back away provisions can cause the company severe economic hardship. The Operator has perhaps looked beyond the prospect area, and considered exogenous variables- how have other Operators in similar prospects fared?

If the economic analysis has considered fields sizes with upside reserves of > 1 billion barrel, the analyses is suspect. An alternative analysis is to examine the chance of success and the size of fields discovered in "analogous" areas. I put the term analogous in quotes, because I know how difficult it is to consider an different geological area as being analogous.

PSC CONSIDERATIONS/CREATIVITY AT THE BARGAINING

Signature bonuses and royalties can be demanded by the host country. The host country is, in effect, ensuring itself against failure, but this can reduce competition for bids. Certain provisions of a PSC may be modified through negotiations. The host country has multiple objectives. The PSC typically has extension/renewal provisions, however if the Operator is particularly successful, the NOC is highly motivated to take over operations.

A PSC should contain provisions to train nationals in the eventual transfer to operate the fields. A lengthy transition of several years may be very beneficial to both the Operator and NOC. The Operator is generally in the best position to continue exploration at the end of a contract. It is incumbent for the Operator to prove to the host country that it will make the NOC a full partner in this effort.

Restoration and abandonment considerations are very important. The problems of Texaco in Ecuador and Occidental in Colombia are excellent examples of why the PSC needs to address environmental concerns associated with R+A. Gas discoveries for offshore fields is an area that I believe will become an increasingly "hot area" for PSC's. The provisions of a PCS must be very carefully written to prevent the national government to demand certain sales price that will deny the Operator's right to a return on their investment.

COMPETITIVE BIDDING

If there are numerous bidder and all are looking at the same information, then the highest bidder is the most optimistic of the group and therefore is the "greater fool" more likely than not to lose money. There is an extensive literature involved in the psychology of competitive bidding, which is outside of the scope of this summary.

3

Water Aquifer Models

INTRODUCTION

Analyses of well test and aquifer pressure responses share a common theoretical basis, the diffusivity equation. Both seek to identify potentially non-unique model parameters. Unlike well testing, the aquifer analyses must work from observational data, very different from a designed test. Also, the time span is commonly in terms of several years. The representation of the aquifer may be used in calculation of original gas or oil in place, or in the forecasting of production.

The model requires assumptions or estimates of the following:

- Geometry- radial, linear, bottom water drive
- Perneability and viscosity (Transmissibility)
- Porosity and compressibility (Storativity)

The irregular shape may be accomodated by either numerical simulation or by using an apparent radius and angle which provide a reasonable equivalent shape. This is discussed in Craft and Hawkins. Similarly, for non-uniform properties, calculation of aquifer volume, and adjusting porosity may be one method. Simulation is another alternative.

We will review three Van Everdingen-Hurst unsteady state models, which are:

- VEH Infinite radial model
- VEH finite radial model
- VEH Linear model

There are other forms including the bottom water drive, which we hope to add later. Additionally, the Carter-Tracy model is used in reservoir simulation. Fetkovich method is an approximation to VEH, with the assumption of pseudo steady-state and was easier to hand calculate. However, today when material balance is done on the computer, the need for simplified solutions is less.

CALCULATION OF CUMULATIVE WATER INFLUX

The starting point is really the conservation of mass equation and Darcy's equation, which result in the diffusivity equation as presented below, and made dimensionless by pressure, time and radius linear transforms..

$$\frac{\partial^2 p}{\partial r^2}+\frac{1}{r}\frac{\partial p}{\partial r}=\frac{\phi\mu c_t}{0.0002637\cdot k}\frac{\partial p}{\partial t}\text{ or}:\frac{\partial^2 p_D}{\partial r_D^2}+\frac{1}{r_D}\frac{\partial p_D}{\partial r_D}=\frac{\partial p_D}{\partial t_D}$$

where time is in hours. Use 0.00633 as constant if time is in days. Dimensionless time, pressure and radius with t in terms of days.

$$t_D=\frac{0.00633\cdot kt}{\phi\mu c_t r_r^2}$$

For constant rate conditions at reservoir/aquifier boundary:

$$P_D=\frac{0.00708\cdot kh(pi-p)}{q\mu}$$

For constant terminal pressure conditons,

$$P_D=\frac{P_i-P}{P_i-P_\gamma}$$

Dimensionless radius is:

$$r_D=r/r_\gamma$$

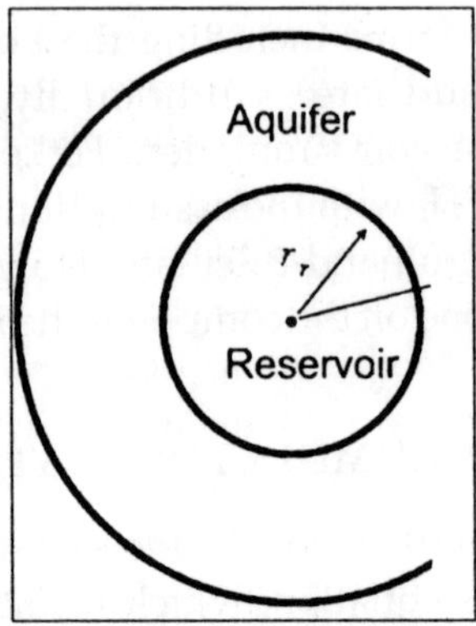

Fig. 3.1 Linear Aquifer Model

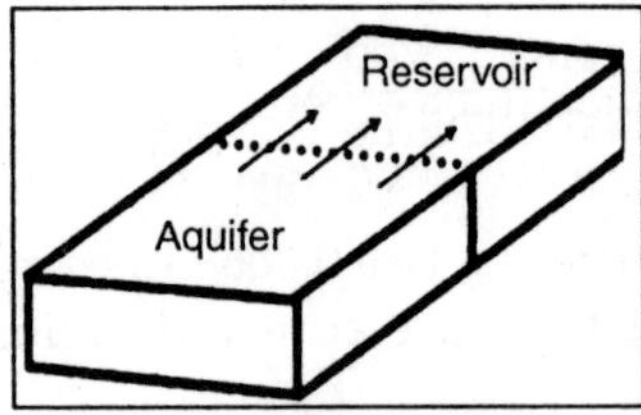

Fig. 3.2

At their best, models can be used to help decision makers evaluate large, complex problems and determine an appropriate course of action. There are many different types of models. Simple physical models such as an aquifer built in an aquarium can help us understand how aquifers function and how they react to different things like rainfall or pollution. Virtual digital simulation models use computers to solve complicated mathematical equations, and they can allow water managers to explore the impacts of various conditions or management plans without actually doing anything that might reduce spring flows or harm endangered species.

Complex digital computer simulations can also help us visualize the aquifer in three dimensions and understand how it works. Models may also take the form of financial spreadsheets that analyse costs and benefits. We are going to look at all these types of models here. Models are usually just one piece of a larger puzzle when we are looking at real-world problems from a systems analysis approach.

Systems analysis is a philosophy of problem solving most often used with large, complex problems where a decision maker collects information and uses everything available to decide what to do. It is not a cookbook approach or a box where we insert data, turn a crank, and get an answer. There are basically three types of questions that modeling addresses, and systems analysis usually involves solution processes and modeling designed to get answers to questions of the third type:

WHY DOES SOMETHING HAPPEN

What is the cause for an effect? Why do springflows increase after it rains? Why do some wells produce more water or different types of water than others? Models of the physical structure of the Edwards can help answer questions like these.

WHAT WILL HAPPEN

Sometimes when we try to answer questions of this type, we can look at past events and make inferences and conclusions about what will happen in the future. For example, we know a lot about what will happen at the J17 monitoring well after it rains. However, most of the Type 2 questions for the Edwards involve things that have never happened before.

If large areas of recharge zone are developed or paved over, what will be the effect on the quantity and quality of recharge water entering the aquifer? What will happen if we draw the level of the Edwards down lower than it has ever been? Digital simulation models can help us answer questions like these.

WHAT IS THE BEST WAY TO DO SOMETHING

For example, how do we manage the Edwards Aquifer? Questions of this type assume that we know what we mean by "best" and that we can measure it. Type 3 questions involve things that have to be balanced in some way like cost, safety, environmental impact, social impact, natural resource conservation, and politics. Society has to decide what the tradeoffs are and what it is willing to accept.

For example, it may be possible to have no environmental impact at a very high cost or lots of environmental impact at a low cost. Society has to decide which one it wants. Also, time is involved when we try to determine what is "best".....is it best for the short term, intermediate, or long term? And, what should go into "do something"? Mandatory rules, voluntary restrictions, or what?

Obviously, questions of the third type are the most difficult to answer. Most of the models that have been developed for the Edwards Aquifer mainly help to answer the first two questions; but they can be used in conjunction with other systems analysis methods to help answer that difficult third question. So far, none of the Edwards models have really been widely used as management tools. Most of the existing models are very complicated and difficult to use in everyday decision-making and Aquifer management.

There has always been concern that models are not calibrated well enough to what really happened in the past. Decision makers want to be certain the models are as fully developed and as accurate as they need to be before using them to make critical decisions. In addition, there are people who don't like what the models suggest! For example, all the models that predict springflow tell basically the same story... reductions in pumping can increase springflows but probably won't keep them flowing in an extreme drought.

One model goes a step further and points out the economic cost of reducing pumpage is huge, yet springs still go dry. And the bottom line is that no one really has a well developed, integrated surface-ground water hydrology model that can be used as a conjunctive management tool. Another problem is that no one really has a good handle on what pumpage has been historically for irrigation and industry.

Municipal pumpage is easy to determine, but when it comes to farming there's a lot of variables and guesswork involved. Still, decisions have to be made, and making decisions in the face of uncertainty, based on incomplete information but using the best data you have, is the stuff of systems analysis.

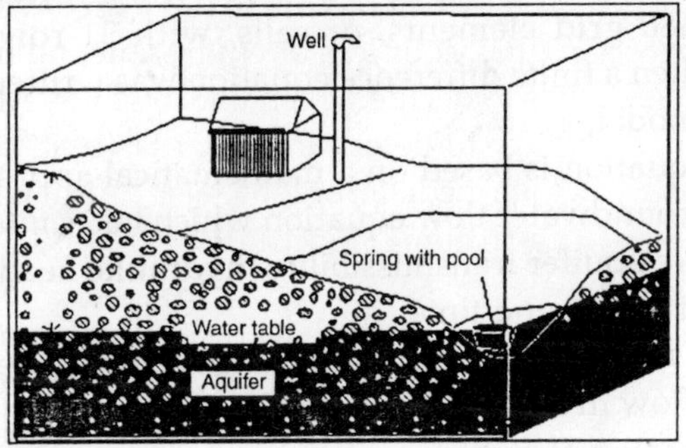

Fig. 3.3

The Bureau of Economic Geology's instructions on how to build an aquifer in a tank! A physical model like this is very helpful in understanding how aquifers function and how they react to different things like rainfall or pollution.

TEXAS WATER DEVELOPMENT BOARD FINITE-DIFFERENCE MODEL

In the mid-1970's, researchers at the Texas Water Development Board evaluated and synthesized previously compiled geologic and hydrologic data, collected additional data, and initiated synthesis and analysis studies using digital computer modeling techniques. These pioneers were William B. Klemt, Tommy R. Knowles, Glenward R. Elder, and Thomas W ·Sieh.

Their objectives were to investigate the influence of projected water demands on Aquifer storage and spring flow, to determine the effectiveness of artificial recharge, and to discover whether ground-water management could protect spring flow at Comal and San Marcos Springs. Their model simulates water levels and spring flows based on the physical constraints of the system and on the recharge and pumpage rates for the Aquifer. Equation for nonsteady flow in a nonhomogenous aquifer

BUILDING THE MODEL

The technique used involved dividing the Aquifer region up into 2,480 grid elements, or cells, with 31 rows and 80 columns. Then a finite difference equation was written for each cell in the model.

Each equation is based on a mathematical approximation of a basic groundwater flow equation which takes into account variables like aquifer transmissibility, hydraulic head, Aquifer storage coefficient, and time.

This is not a math lesson, but the governing equation for nonsteady flow in a nonhomogenous Aquifer was a derivation of the one shown at right.

$$\frac{\partial}{\partial x}[T\frac{\partial h}{\partial x}]+\frac{\partial}{\partial_Y}[T\frac{\partial h}{\partial_Y}]-S\frac{\partial h}{\partial t}+Q,$$

where

T = aquifer transmissibility (Length squarred/Time):
h = hydraulic head (Length)
S – aquifer storage coefficent;
t – time (Time);
Q = net ground-water flux per unit area (Length/ Time);
x,y – rectangular coordinates (Length).

There are three types of cells in the model: outcrop cells, artesian cells, and boundary cells. In the graphic the boundary cells are the ones with at least one black, bold border; the outcrop cells are in dark grey; and the artesian cells are in light gray. Most of the grid cells lie outside the boundary, so only 856 of the cells in the grid were considered as part of the Edwards system and included in the simulation.

Critical steps in model building are the assignment of recharge and pumpage values to each of the cells. Recharge was divided into two components: water that recharges by going down through sinkholes and faults in creek beds, and water that becomes recharge by falling directly onto the outcrop. In the assignment of recharge, the amount of water known to have been lost from a stream to recharge was assigned evenly to the cells that contained the stream reach, and the amount recharged by direct infiltration was evenly divided among the remaining cells for that subbasin.

Assignment of pumpage is also critical - it must be assigned so it approximates the distribution of pumpage that actually occurred. In this model, the pumpage assigned to each cell is the total pumpage of all wells located within the cell.

MODEL CALIBRATION

After a model is initially built, there comes a calibration phase that involves comparing observed and simulated water levels. The idea is to find out how well the model can predict what actually happened in the past. For this model, the calibration period used data from 1947 to 1971. This period was chosen because the data was readily available and these years included a great variation in water levels.

There was an extreme drought that lasted from 1947 until 1956, and 1957 and 1958 were years of unusually high recharge. It was felt that if the model could simulate water levels for this period, it would be well calibrated.

The difference amounts to only 4.3 per cent of the total flow. Based on this comparison, it was decided the digital model was calibrated to a degree of accuracy sufficient to reproduce past events and that the model could therefore be used to predict future.

FUTURE SIMULATION PHASE

After calibration, a Future Simulation Phase involved using the model to predict what would happen in the Aquifer under a variety of different conditions. There were a total of 20 model runs that looked at the response of the Aquifer under a variety of pumpage and recharge conditions, various drought and flood conditions, several Aquifer management plans, and under augmentation pumping.

By the year 2000, total Aquifer pumpage was projected to be around 650,000 acre-feet. Comal Springswas projected to go dry in 1995, and San Marcos Springs would be dry by 2010. Pumpage has not really panned out as was projected back in the 70's when this model was built. Currently there is a cap on pumpage at 450,000 acre feet, and the Springs are still flowing.

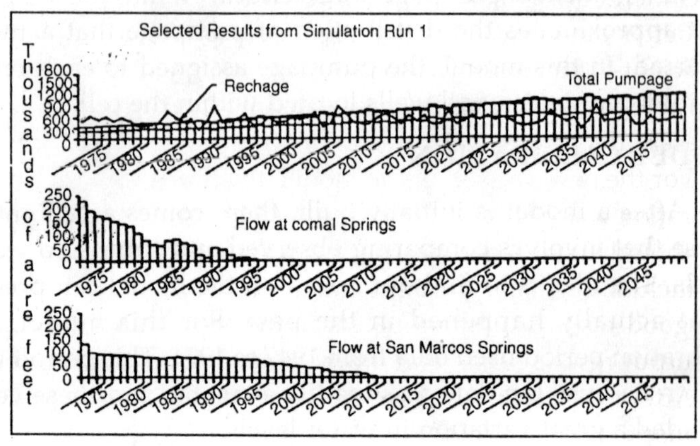

Fig. 3.4

MODEL CONCLUSIONS

The 20 model runs suggested the Aquifer could indeed meet the demand projections, but the natural flow of Comal and San Marcos Springs would cease by 2020 if projected pumpage occurred as predicted. It also concluded the addition of artificial recharge that was being proposed at the time did not result in an appreciable increase in the Aquifer's available water. Further, it suggested the drought-flood sequence of the 1950's produced only minor effects compared to the long-term influence of pumpage by humans.

The model also concluded that spring flows at Comal and San Marcos Aprings could be maintained through groundwater management plans. Finally, it suggested there is sufficient storage in the Aquifer to allow spring flow of Comal and San Marcos

1992 MODEL REFINEMENT

In 1992, David Thorkildsen and Paul D. McElhaney revisited the 1970's model, refined it, and applied it under new conditions of reduced regional pumpage and under new Aquifer management plans. During the years between 1979 and 1992, the United States Geological Survey continued to collect new and better data on things like Aquifer transmissivity, storage coefficients, and anistropy, so these two geologists at the TWDB were able to incorporate the new data into their model to better represent the flow system and improve the model simulation.

For the new version of the model, the time period used for calibration was 1947 to 1959 instead of 1947 to 1971 as before. It was thought that since this period was one of extreme stress that included a major drought and a recovery period, it would be a good test for the parameters in the model. The time step for output results became monthly instead of annually.

After calibration, the model was used to simulate Aquifer response to several constant pumpage scenarios under recharge conditions of the drought and recovery sequence. The results indicated that large pumpage reductions would be necessary in order to ensure minimum springflows in the 40 to 50 cfs range at both San Marcos and Comal Springs. For a flow of 50 cfs or greater to be maintained at Comal Springs, the model indicated total pumpage would have to be in the 225,000 to 250,000 acre-feet per year range for the entire San Antonio region. This is much less than current pumpage.

To maintain 100 cfs or greater, the model indicated maximum total pumpage would be about 165,000 acre feet per year. The model also looked at the effect of reduced pumpage in the western, agricultural part of the Edwards region. Interestingly, the model indicated this would result in an increase in springflow at Comal Springs, but the increase was not equal to the amount of pumpage reduced. If the reduction in pumpage was placed in Uvalde county, only 34% of the reduction shows up as springflow at Comal Springs after nine years.

If the reduction was placed in Medina county, 67% of the reduction was seen as increased springflow at Comal Springs after nine years. In both cases the remainder of the reduced pumpage either became increased Aquifer storage, interformational flow, or increased outflow. When the reductions were placed in Uvalde county, Leona springs began to flow again; but not when the reductions were placed in Medina county.

One of the ideas being floated in the early 90's by the Texas Water Commission was a mandatory water curtailment plan called the "dry-year option", under which pumpage for municipal, industrial, and aquaculture was adjusted (decreased or increased) on a monthly basis based on the simulated water level in the model cell representing the J17 monitoring well. Irrigation pumpage was to be adjusted on an annual basis, also using the simulated water level at J17.

Under this plan, the model indicated that San Marcos Springs would always flow, but Comal Springs would cease to flow in an extreme drought.

UTSA'S DECISION SUPPORT SYSTEM

In the mid-90's the University of Texas at San Antonio built something called the Decision Support System for the Edwards. It was actually the TWDB model converted to run on a personal computer, with minor embellishments and additions. This was the first model that had an interface sufficiently developed that a water manager could use it behind a desk. It has not been widely used outside academic circles.

UNITED STATES GEOLOGICAL SURVEY FINITE DIFFERENCE MODEL

In 1988, Robert.W. Maclay and L.F. Land, working for the United States Geological Survey, developed an Edwards model that focused on concepts such as the effects of barrier faults and storage within the aquifer. Like the TWDB model, it was a finite-difference model, but it was modified to provide the capability of representing barrier faults.

Rather than predicting Aquifer levels under different conditions as in the TWDB model, the main objective was to learn more about storage of water in different parts of the Aquifer and identification of flowpaths within the aquifer.

FINITE DIFFERENCE GRID FOR USGS MODEL

More than 300 simulations were performed, and the modeling effort resulted in several major interpretations. One significant interpretation was that two essentially independent areas of regional flow exist in the west and central portions of the Aquifer, and flows from the two areas converge at Comal springs.

Barrier faults deflect patterns of groundwater movement, and the most noticeable deflection is the convergence of flow through a geologic structural opening, the Knippa Gap, in eastern Uvalde county.

A second significant interpretation is that groundwater flow in northeastern Bexar, Comal, and Hays counties is diverted by barrier faults towards San Marcos Springs. The simulations showed that several barrier faults in the northwestern part of the San Antonio area had a significant effect on storage, water levels, and springflow within the Edwards Aquifer.

LUMPED PARAMETER MODEL

A lumped parameter model for the Edwards was developed in 1992 by Nisai Wanakule and Roberto Anaya with funding from various agencies. Their objective was to develop and test a simple and easy to use model to accurately simulate monthly water levels and springflows of the Edwards Aquifer. The model was formulated using a discrete, nonlinear, non-stationary system based on control theory. The main advantage of this simulation model was that it could utilize limited data to provide accurate, short-term estimates of Aquifer levels and springflows.

Previous models required massive amounts of inputs and generated only annual estimates. Basically, the Edwards Aquifer was represented as a series of connected rock-filled tanks that represent eight major drainage basins of the Aquifer and springflows at Comal and San Marcos springs. Individual drainage basins included the Nueces River, the Frio River, the Sabinal River, Upper and Lower Seco and Hondo Creek the Medina River, Helotes and Salado Creek, Cibolo Creek, and the

Blanco River. Data from the Guadalupe River basin were not available and thus were not included. Inputs to the model included real time measurements of recharge from river gauging stations. Outputs of the model included forecasts of levels in seven key observation wells and springflows. The model estimates monthly recharge, water levels, and pumping in each basin. It also allows for the interaction of ground and surface water between sub-basins.

Monthly stream gauge and Aquifer level data from 1975 to 1990 were used to calibrate the model, while data from 1962 to 1974 were analysed to verify its performance. The simulation exercises showed the model is accurate and very easy to use because much less data is needed. When a Macintosh computer with a 68040 microprocessor was used, the lumped parameter model efficiently simulated 189 monthly iterations of water levels for nine drainage basins in less than four minutes.

Recharge functions in the model provided good estimates of monthly recharge. Recharge estimates were very similar to results from the USGS model in most cases, although peak years were under-represented.

A & M UNIVERSITY ECONOMIC/HYDROLOGIC MODEL

In 1993, Bruce McCarl and Lonnie Jones of the Texas A & M University Agricultural Economics Department worked with graduate students R. Lynn Williams and Carl Dillon to investigate the implications of management plans that were being proposed for the Aquifer at the time. At that time, hardly any research had been done that looked at the effect of different management plans on the regional economy.

In 1992, two management plans were drafted by the Texas Water Commission. The "Sector Plan" proposed specific cutbacks for agricultural, municipal, and industrial users when Aquifer levels at the J17 index well dropped to key levels, and the "Total Limit Plan" called for a total reduction in water use, but didn't specify where the cutbacks had to occur. Neither plan was adopted, but the Texas Legislature began considering plans similar to the TWC plans. A study was undertaken to evaluate the hydrologic and economic implications of the TWC plans. The A & M researchers analysed four variants of the plans using an annual economic/hydrologic simulation model. The model simulated water use by the agricultural, industrial and municipal sectors while simultaneously forecasting annual spring flow and year-end water elevations. The model also accounted for uncertainty in recharge amounts and in the incidence of elevation-triggered pumping limits.

A key feature of the model was that it optimized optimize water allocations among the various sectors of users based on maximizing economic welfare. In other words, water was allocated to the activities that generated the greatest economic value. The model suggested that the simultaneous imposition of pumpage limits, water rights and water markets would be necessary to maintain economic efficiency. To allocate water towards the highest value use, the model made the assumption that low valued users would allow high value user to use their water instead. In reality, this would be highly unlikely without compensation for the low valued users.

So, a rights structure was also examined where the irrigated agriculture sector (generally a low valued user) was guaranteed water usage at their 1988 level. The model showed that under these circumstances, there are gains in agricultural welfare, but municipal, industrial and total welfare is reduced by more than the agricultural gain. And without an agricultural guarantee for an allocation at their 1988 level, the municipal and industrial gains are achieved at agriculture's expense.

Either way, the model suggested everyone loses out economically when pumping is no longer unrestricted and water is allocated. It indicated the management plans were likely to result in annual losses for current water users ranging from $0.73 million to $1.5 million. When the value of water is optimized in an economic sense, the model predicted an annual loss in regional economic activity of between $6.26 and $19.58 per acre foot that pumpage is reduced. By the year 2000, agricultural use value was equal to about $19 per acre foot while non-agricultural values were about $109 per acre foot. This suggested that paying farmers not to use water would make economic sense. The benefits of the management plans that were being proposed were increased flows at Comal and San Marcos Springs and higher year-end Aquifer levels. This model was the first attempt to quantify the economic cost of keeping springs flowing. It suggested costs would be very high, and there would still be no guarantee that Comal Springs would keep flowing in a severe drought.

UNITED STATES GEOLOGICAL SURVEY FINITE-ELEMENT MODEL

In 1994, Eve Kuniansky and Kelly Holligan, working for the United States Geological Survey, published a model that sought to understand and describe the regional flow system and examine how the Edwards Aquifer is connected to the Trinity Aquifer and other hydraulically connected units. The researchers wanted to gain a better understanding of how much water enters the Edwards from other Aquifers and how anistropy affects regional flows of groundwater. Anistropy occurs when water can move more easily through the rock matrix in one direction than another.

In most Aquifers made up of flat-lying sedimentary rocks, water can move more easily horizontally than vertically. Because of this property, and because the Aquifer is many times wider than it is deep, groundwater flow through the Aquifer can be approximated with two dimensions even though the Aquifer actually exists in three dimensions. Unlike the previous models that used a finite-difference approach, this model used a finite-element method because it is one of the few numerical methods that can approximate anistropy.

The finite element method of solving the flow equation differs from the finite difference approach in that it involves piecewise approximate of the flow domain instead of a regular grid. The flow domain is broken into discrete subdomains, called finite elements. The simplest element is triangular with linear sides.

This mesh was designed on the basis of surface water drainage divides and streams, transmissivity subregions, and geographic subareas. Notice how this mesh covers the Edwards aquifer region covered by the other models, and also includes almost all the Texas Hill County and portions of west and northwest Texas.

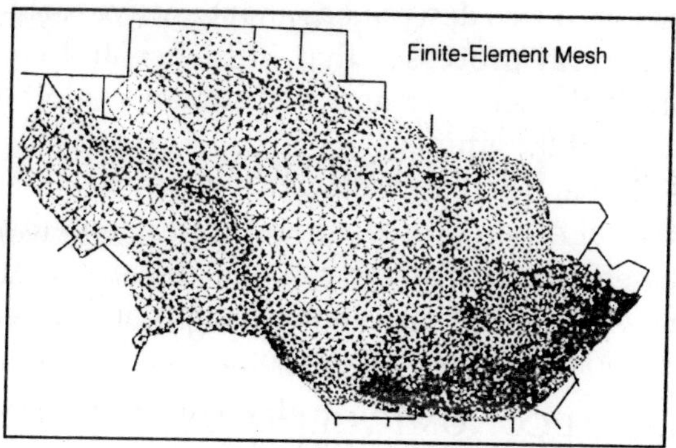

Fig. 3.5

In this model, the major departure from the earlier simulations was this one allowed a hydraulic connection from the Trinity Aquifer. The previous models assumed a no-flow boundary between the Edwards and the Trinity. Two simulations were performed.

One was for the winter of 1974-1975 because there was adequate water-level data for model calibration for that period; and the other simulation was for predevelopment conditions.

The simulations indicated that a significant amount of water enters the Balcones Fault Zone from other aquifers; about 97,500 acre-ft per from the Trinity and an additional 292,500 acre-ft per year from the Edwards-Trinity Plateau aquifer west of the Haby Crossing fault.

Also, large irrigation withdrawals from the Edwards-Trinity Aquifer and the Cenozoic Pecos alluvial Aquifer in the Trans Pecos region have resulted in the drying up of several springs including Comanche and Leon Springs. Further, the results also suggested that anistropy is an important factor in groundwater movement in the Balcones Fault Zone.

EDWARDS AQUIFER SIMULATION

In 1995 the Edwards Aquifer Simulation (EAS) was developed under the guidance of the Edwards Underground Water District. It was designed as an interactive tool to explore both the natural processes that influence and control the operation of the Aquifer and the human exploitation and management of the aquifer.

It used a hands-on, role-playing environment that allowed users to model Aquifer response and springflows under many different scenarios such as varying rainfall patterns, institution of pumping restrictions, allocation of supply, and establishment of conservation and reuse programmes.

BUREAU OF ECONOMIC GEOLOGY 3D VIRTUAL REALITY MODEL

The Bureau of Economic Geology recently unveiled theirAdventures in Virtual Reality that includes an awesome 3D virtual reality tour of the Edwards Aquifer. You can rotate the model and experiment with different viewpoints, zoom in for a closer look, and look at the locations of wells and cities.

3D MODEL OF THE EDWARDS IN NORTHERN BEXAR COUNTY

In 2004 Michael P. Pantea and James C. Cole of the USGS completed a digital, three-dimensional faulted hydro-stratigraphic model to represent the geologic framework of the Edwards aquifer system in the northern Bexar county.

The model is based on mapped geologic relationships that reflect the complex structures of the Balcones fault zone, detailed lithologic descriptions and interpretations of about 40 principal wells (and qualified data from numerous other wells), and a conceptual model of the gross geometry of the Edwards Group units derived from prior interpretations of depositional environments and paleogeography.

HSPF RECHARGE MODEL FOR THE SAN ANTONIO SEGMENT OF THE EDWARDS

In 2005 LB-Guyton Associates, working for the Edwards Aquifer Authority, published a report on their development of nine models for basins that recharge the San Antonio segment of the Edwards. They used HSPF a popular and flexible modeling tool first developed in the early 1960s that simulates the hydrologic processes, and associated water quality, on pervious and impervious land surfaces and in streams and well-mixed impoundments.

HSPF can be adapted for almost any modeling task, and it contains hundreds of process algorithms developed from theory, laboratory experiments, and empirical relations from instrumented watersheds. The most productive and easiest way to run HSPF is from a DOS prompt, using structured input and output text files, so this model is pretty much for computer geeks only.

For the Edwards model, the main objective was to estimate historical recharge from 1950 to 2000. Traditionally, the USGS has used a water balance method to calculate recharge that relies on measuring total monthly flows in gauges upstream and downstream of the recharge zone. It is a very straightforward approach that uses readily available streamflow, precipitation, and curve number data.

A slightly revised version of this approach was used by HDR Engineering in connection with the Trans-Texas water project. The results are generally comparable, although USGS estimates during "wet" years are usually higher than HDR estimates, and HDR estimates are consistently higher in several basins. The models were constructed to incorporate rainfall, evaporation, topography, channel-loss, land use, vegetation, geologic and soil characteristics, and water diversions to simulate the hydrologic processes at an hourly time step.

The model also suggested that the source of recharge varies dramatically among the basins depending on the channel loss characteristics, areal extent of recharge zone, and volume of upstream flows from the contributing zone. In some basins, land recharge is a major component of overall recharge and in other basins recharge is dominated by channel loss. One of the most sensitive parameters in the model was the channel loss in streambeds over the recharge zone. There were no real surprises or worldview-shattering revelations in the HSPF model results, but a secondary objective of the project was to develop models that would be compatible with future assessments of possible initiatives such as brush control, recharge dams, flood control projects, and weather modification.

4

Reservoir Models

MODELS – RADIAL

Rectangular reservoir with a vertical well located anywhere inside.

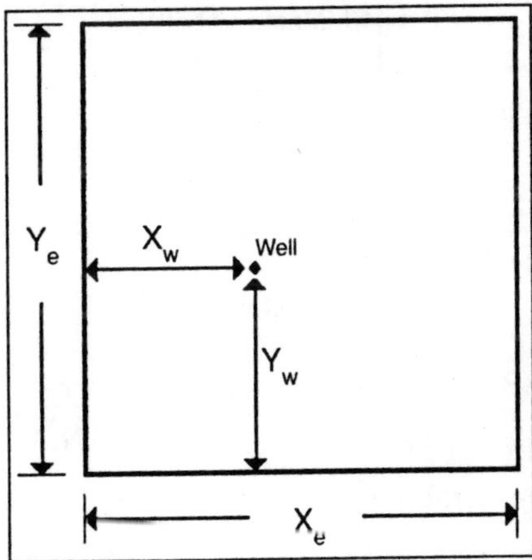

Fig. 4.1

The Radial Model is actually a homogeneous rectangular reservoir model, with a vertical well located anywhere inside. The model is analytical and uses Green's functions and method of images to calculate a fully integrated wellbore pressure history from early-time, through transition and into boundary dominated flow. The model incorporates storage and skin effects. Rate dependent (turbulence) skin can also be accommodated (using a semi-analytical method) by entering a "D" turbulence factor.

The x and y dimensions of the reservoir can be set to any value above zero, and location of the well may be set to anywhere within the reservoir boundaries. Pseudo-time is used for gas wells, to accommodate changing gas compressibility with reservoir pressure.

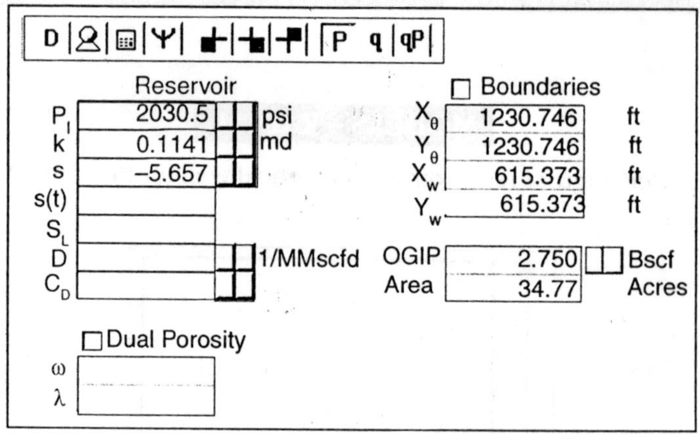

Fig. 4.2

PARAMETERS

- Pi = initial pressure
- k = permeability
- h = net pay
- s = skin (may be left blank)
- D = turbulence factor (may be left blank)
- C_D = storage coefficient (may be left blank)
- x_e = reservoir length
- y_e = reservoir width
- x_w = x position of well within reservoir
- y_w = y position of well within reservoir

- O_IP = original gas/water/oil in place
- Area = extent of reservoir

MODELS – HORIZONTAL

Rectangular reservoir with a horizontal well located anywhere inside.

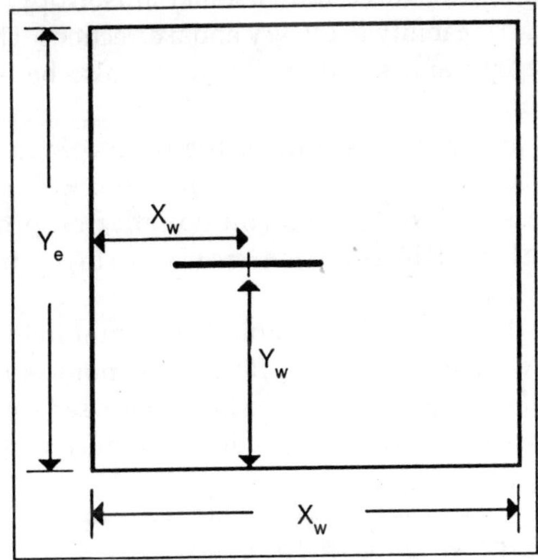

Fig. 4.3

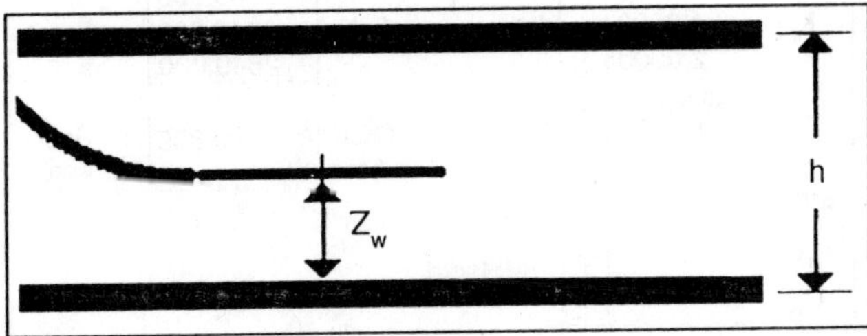

Fig. 4.4

The Horizontal Well Model is a homogeneous rectangular reservoir model, with a horizontal well located anywhere inside. The model is analytical and uses Green's functions and method of images to calculate a fully integrated wellbore pressure history from early-time, through transition and into boundary dominated flow.

The model incorporates storage and skin effects. Rate dependent (turbulence) skin can also be accommodated (using a semi-analytical method) by entering a "D" turbulence factor. The model allows three dimensional anisotropy, providing inputs for permeability in the x, y and z directions. The effective wellbore length and stand-off height are also be included as model inputs.

The x and y dimensions of the reservoir can be set to any value above zero, and location of the well may be set to anywhere within the reservoir boundaries, provided the horizontal wellbore does not overlap any of the boundaries.

Thus, it is possible to model the combined effects of horizontal wells and boundary flow in any number of different geometrical configurations. Pseudo-time is used for gas wells, to accommodate changing gas compressibility with reservoir pressure.

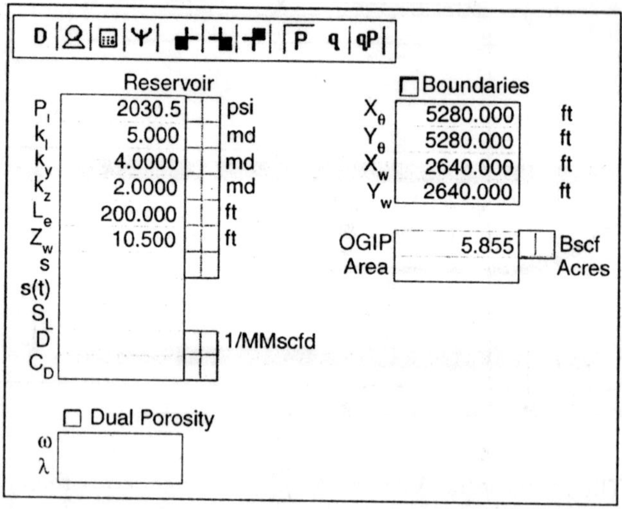

Fig. 4.5

PARAMETERS

- P_i = initial pressure
- k_x, k_y, k_z = permeability in the x, y and z directions
- x_f = fracture half-length
- L_e = effective wellbore length
- z_w = z position of horizontal well within reservoir
- s = skin
- D = turbulence factor (also known as non-Darcy coefficient or turbulence skin effect)
- C_D = wellbore storage coefficient
- x_e = reservoir length
- y_e = reservoir width
- x_w = x position of well within reservoir
- y_w = y position of well within reservoir
- O_IP – original gas/water/oil in place
- Area – extent of reservoir

MODELS – FRACTURE

Rectangular reservoir with a vertical infinite conductivity fracture located anywhere inside.

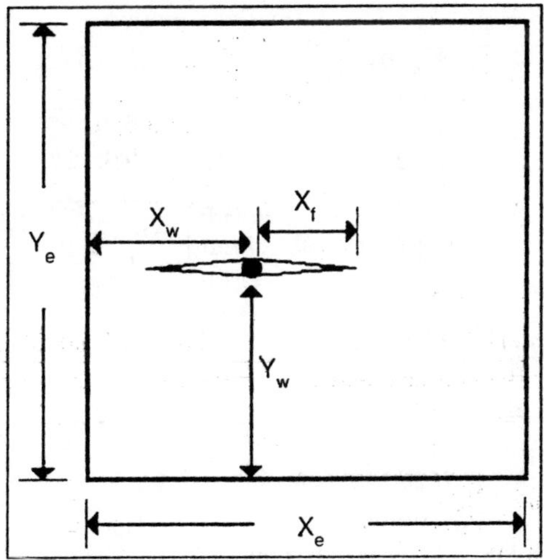

Fig. 4.6

The Fracture Model is a homogeneous rectangular reservoir model, with a vertical hydraulic fracture (of infinite conductivity) located anywhere inside. The model is analytical and uses Green's functions and method of images to calculate a fully integrated wellbore pressure history from early-time, through transition and into boundary dominated flow.

The model incorporates storage effects. Fracture half-length and fracture face skin may be specified by the user, also. Rate dependent (turbulence) skin can also be accommodated (using a semi-analytical method) by entering a "D" turbulence factor.

The x and y dimensions of the reservoir can be set to any value above zero, and location of the well may be set to anywhere within the reservoir boundaries, provided the fracture does not overlap any of the boundaries.

Thus, it is possible to model the combined effects of fracture and boundary flow in any number of different geometrical configurations. Pseudo-time is used for gas wells, to accommodate changing gas compressibility with reservoir pressure.

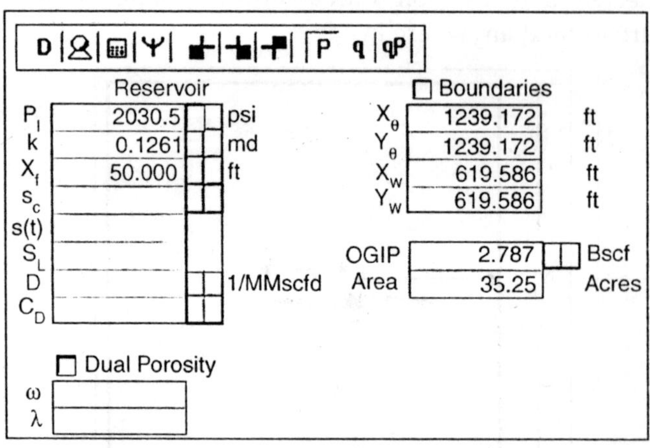

Fig. 4.7

PARAMETERS

- P_i = initial pressure

- k = permeability
- x_f = fracture half-length
- sc = choke skin (may be left blank)
- D = turbulence factor (may be left blank)
- C_D = storage coefficient (may be left blank)
- x_e = reservoir length
- y_e = reservoir width
- x_w = x position of well within reservoir
- y_w = y position of well within reservoir
- O_IP = original gas/water/oil in place
- Area = extent of reservoir

MODELS – COMPOSITE

A circular reservoir formed of concentric regions with a vertical well located at the center. The following schematic shows three regions, each with their own net pay value.

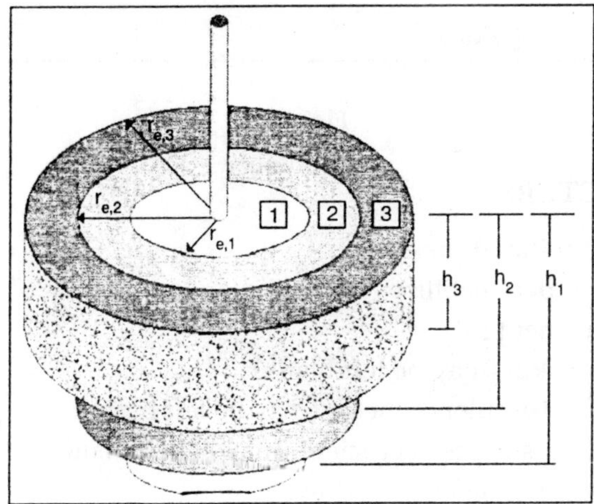

Fig. 4.8

The composite model is composed of homogeneous concentric rings, with a vertical well located at the center. The model is analytical and calculates a continuous sandface pressure history from early-time, through transition and into boundary dominated flow. The model incorporates storage and skin effects. Rate dependent (turbulence) skin can also be accommodated (using a semi-analytical method) by entering a "D" turbulence factor.

The outer radius (r_o) for each ring can be set to any value that is greater than the r_o value for an inner ring. It is also possible to remove regions until all that remains is one region. Each region can have its own net pay (h). Pseudo-time is used for gas wells, to accommodate changing gas compressibility with reservoir pressure.

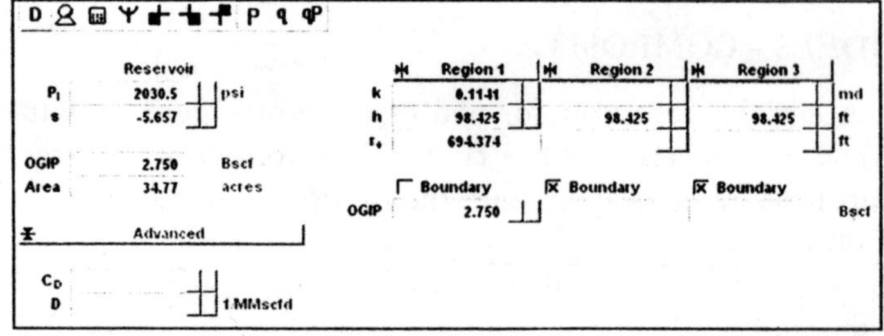

Fig. 4.9

PARAMETERS

- P_i = initial pressure
- k = permeability
- h = net pay
- s = skin (may be left blank)
- D = turbulence factor (may be left blank)
- C_D = storage coefficient (may be left blank)
- r_o = region outer radius
- O_IP = original gas/water/oil in place
- zArea = extent of reservoir

MODELS – MULTILAYER

A circular reservoir formed of stacked layers with a vertical well located at the center.

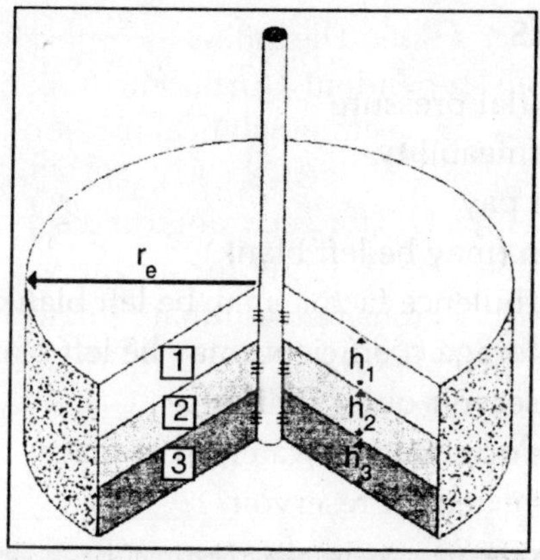

Fig. 4.10

The multilayer model is composed of cylindrical regions that are stacked vertically. The model is analytical and calculates a coı.tinuous sandface pressure history from early-time, through transition and into boundary dominated flow.

The model incorporates storage and skin effects. Rate dependent (turbulence) skin can also be accommodated (using a semi-analytical method) by entering a "D" turbulence factor. The outer radius (r_o) for each ring can be set to any value above zero. It is also possible to all the layers until all that remains is one layer. Each region can also have its own net pay (h). Pseudo-time is used for gas wells, to accommodate changing gas compressibility with reservoir pressure.

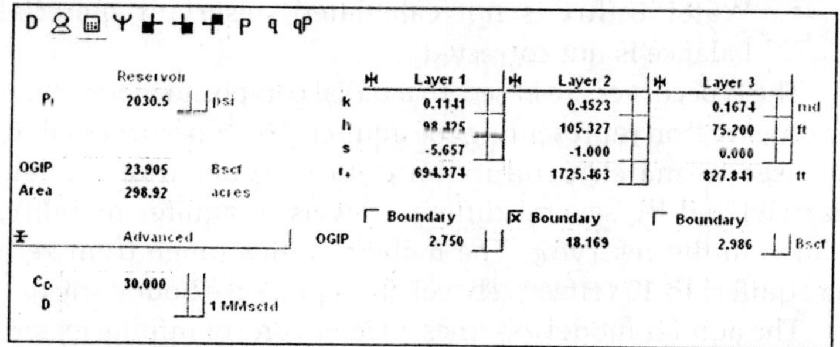

Fig. 4.11

PARAMETERS

- P_i = initial pressure
- k = permeability
- h = net pay
- s = skin (may be left blank)
- D = turbulence factor (may be left blank)
- C_D = storage coefficient (may be left blank)
- r_o = reservoir outer radius
- O_IP = original gas/water/oil in place
- Area = extent of reservoir

WATER DRIVE (SINGLE PHASE)

TRANSIENT WATER DRIVE MODEL

The Transient Water Drive Model consists of dimensionless typecurves, implemented in Blasingame, Agarwal-Gardner and NPI. The typecurve model has the following characteristics:

- Assumes an infinite acting aquifer with a stationary boundary
- Assumes a radial edge water drive system (radial composite model)
- Ideal for very large aquifers of low to medium mobility
- Designed for Early-Time analysis, as it is a Pressure Transient model
- Water influx is not calculated, reservoir material balance is not corrected

The typecurves are based on a radial composite model, with the outer region representing the aquifer. The typecurves allow the user to match production data to typecurve models characterized by several different levels of aquifer mobility, relative to the reservoir. The mobility ratios range from zero (no aquifer) to 10 (effectively constant pressure boundaries).

The aquifer model assumes a water zone of infinite extent. Thus, matching to the water-drive typecurves will allow the user to determine the oil/gas –in-place, as well as "aquifer strength" in the form of a mobility ratio (the aquifer permeability is also calculated based on an assumed water viscosity).

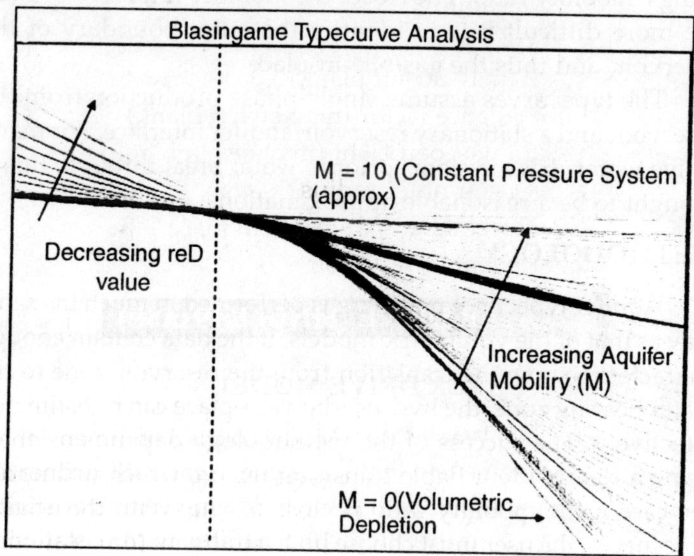

Fig. 4.12 The Dimensonless Type Curves and Definitions

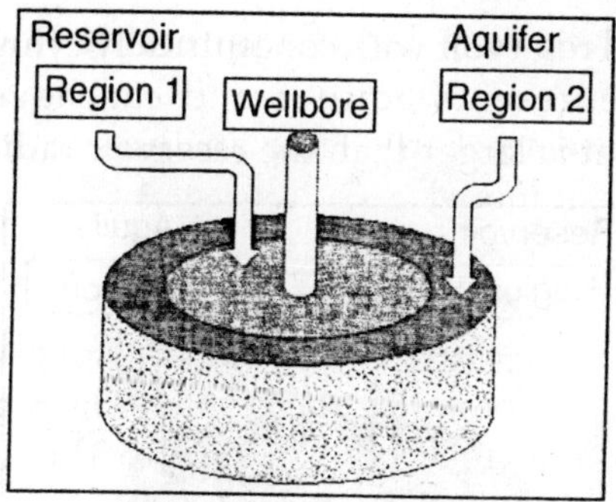

Fig. 4.13 Mobility Ratio (M)

$$M = \frac{M_{aq}}{M_{res}} = \frac{k_{aq}}{k_{res}} \frac{\mu_{res}}{\mu_{aq}}$$

The practicality of estimating hydrocarbon volumes using water-drive typecurves is a strong function of, among other things, mobility ratio. The closer the mobility ratio is to one (1), the more difficult it is to identify the outer boundary of the reservoir, and thus the gas/oil –in-place

The typecurves assume single-phase production from the reservoir, and a stationary reservoir/aquifer interface. For many active water-drive systems prior to water breakthrough, this is thought to be a reasonable approximation.

METHODOLOGY

Aquifer typecurve matching is performed in much the same way as that of the volumetric models. If the data contain enough character to show the transition from the reservoir zone to the water bearing zone, the hydrocarbons-in-place can be estimated effectively. The success of the methodology depends on there being a clearly identifiable transition region, which will not be the case if the mobility ratio is close to one. With the aquifer typecurves, the user must choose both a transient (dimensionless reservoir radius) stem and an Aquifer Mobility stem.

MODELS - WATER DRIVE

Cylindrical reservoir with concentrically cylindrical aquifer. Outer zone represents aquifer and can have any radius (provided that it is larger than the reservoir radius).

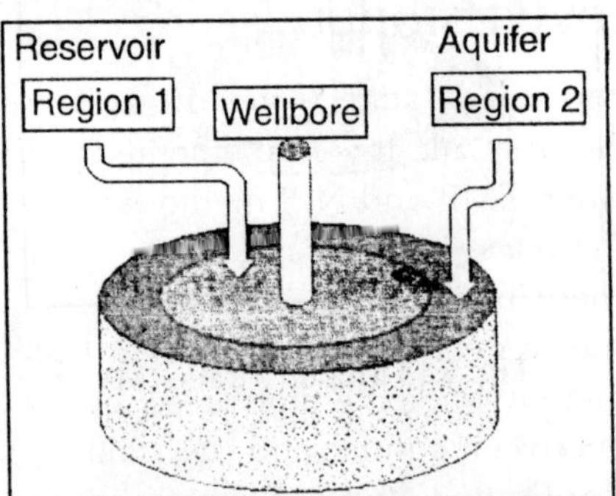

Fig.4.14 Edge Water Drive Model (Composite Model)

The Water Drive Model is an analytical radial composite reservoir model. The reservoir is represented by the inner zone, and is given a radius of "re". The outer zone represents the aquifer, with a radius of "r_{aq}". The aquifer radius may be set to any value greater than "r_e". The model accounts for the mobility difference between reservoir and aquifer by allowing the user to enter a mobility ratio (M). This is defined as the ratio of the aquifer mobility to that of the reservoir.

$$M = \frac{M_{aq}}{M_{res}} = \frac{k_{aq}}{k_{res}} \frac{\mu_{res}}{\mu_{aq}}$$

PARAMETERS

- P_I – initial reservoir pressure
- k - permeability
- s - skin
- D – turbulence factor (also known as non-Darcy coefficient or turbulence skin effect)
- C_D – wellbore storage constant
- r_e – boundary radius for fluid (inner) zone
- O_IP – original gas/water/oil in place
- M – mobility ratio
- r_{aq} – boundary radius for aquifer (outer) zone

PSS WATER DRIVE MODEL

The pseudo-steady-state water influx model is based on work by Fetkovich et. al.. It is implemented in the Blasingame, Agarwal-Gardner, FMB and NPI methods.

The model characteristics are as follows:

- Assumes finite aquifer, modeled as a tank
- Assumes a geometry independent transfer coefficient (productivity index) which prescribes how much water flows between aquifer and reservoir
- Ideal for limited aquifers of medium to high mobility
- Designed for EARLY-TIME analysis, prior to water break-through

- Water influx is calculated, reservoir material balance includes the effect of water influx

The conventional use of the PSS water drive model is in history matching P/Z data from gas reservoirs. The objective of the model is to simultaneously solve the aquifer influx equation with the material-balance equation for the reservoir.

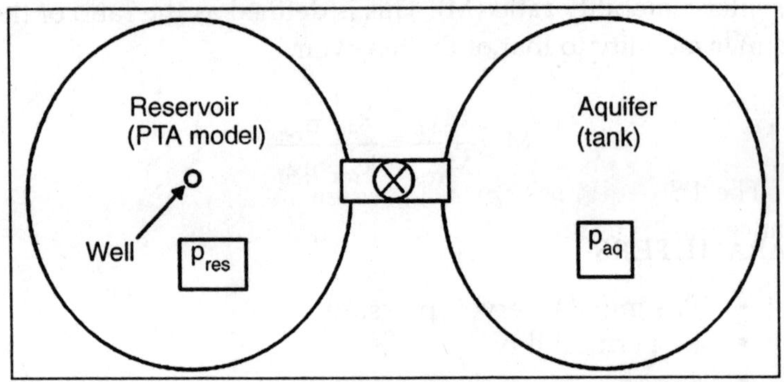

Fig. 4.15

METHODOLOGY

The implementation of the PSS method involves a process called inverse modeling. This process uses the standard (volumetric) typecurves as a base model and *subtracts*the water influx effect from the data, so as to collapse the production response (q/Dp vs tca) back onto the base typecurve.

It is clear that the aquifer effect must be apparent in the data for the inverse modeling procedure to be effective. The key matching parameters in the PSS model are the Initial Water In Place (IWIP) and the Aquifer Productivity Index (PI).

MATERIAL BALANCE TIME FOR WATER DRIVE RESERVOIRS

OIL RESERVOIRS (UNDERSATURATED)

Starting with a general (under-saturated) oil material balance (including water influx):

$$QB_o = NB_{oi}\left(c_o\left(p_i - \bar{p}\right) + \frac{CwSw + Cf}{So}\left(p_i - \bar{p}\right)\right) + W_eB_w$$

Rearranging equation (1) and dividing both sides by q

$$\frac{P_i - \bar{P}}{q} = \frac{QB_o - W_eB_w}{NB_{io}C_eq}$$

Where

$$c_e = c_o + \frac{c_w s_w + c_f}{s_o}$$

The PSS equation for oil, written in terms of material-balance time is as follows:

$$\frac{p_i - \bar{p}}{q} = \frac{t_c}{NC_e}$$

Comparing equations, we see that

$$t_c = \frac{QB_o - W_eB_w}{B_{oi}q}$$

GAS RESERVOIRS

For gas reservoirs, there are two non-linearities that exist:

1. Compressibility factor (Z) and viscosity vs. reservoir pressure
2. Compressibility (c) and viscosity vs. time

In order to use the PSS equation for gas, these non-linearities must be linearized. For reservoir pressure vs. Z, this is done using pseudo-pressure, and for compressibility vs. time, this is done using pseudo-time. Thus, we want to derive a form of the PSS equation that utilizes pseudo-pressure (and pseudo-time (tca). Starting with a general form of the gas material balance (including water influx):

$$\frac{\bar{P}}{\bar{z}} = \frac{\frac{p_i}{z_i}\left(1 - \frac{G_p}{G}\right)}{1 - c_e\left(pi - \bar{p}\right) - Ei\frac{W_e}{G}}$$

Where

$$Ei = \frac{V_{sc}}{V_i} = \frac{pi}{p_{sc}} \frac{T_{sc}}{Ti} \frac{Z_{sc}}{Zi}$$

The definition of pseudo-pressure is:

$$\overline{\psi}(\overline{p}) = 2 \int_0^{\overline{p}} \frac{\overline{p}dp}{\overline{\mu Z}}$$

In order to get to the PSS equation for gas, we can use the chain rule as follows:

$$\frac{\partial \psi}{\partial t} = \frac{\frac{\partial}{\partial t}\left(\frac{\overline{p}}{\overline{z}}\right) * \frac{\partial \psi}{\partial \overline{p}}}{\frac{\partial}{\partial \overline{p}}\left(\frac{\overline{p}}{\overline{z}}\right)}$$

The first term in the numerator is calculated as follows:

$$\frac{\partial}{\partial t}\left(\frac{\overline{p}}{\overline{z}}\right) = -\frac{p_i}{z_i}\frac{q}{G}\left[1 - ce(pi - \overline{p}) - Ei\frac{W_e}{G}\right]^{-1} - \frac{pi}{zi}\left[1 - \frac{Gp}{G}\right]$$

$$\left[ce\frac{\partial}{\partial t} - \frac{Ei}{G}q_{we}\right]\left[1 - c_e(p_i - \overline{p}) - Ei\frac{W_e}{G}\right]^{-2}$$

Recognizing that

$$\frac{\partial \psi}{\partial t} = \frac{\partial \psi}{\partial p}\frac{\partial p}{\partial t}$$

Where

$$\frac{\partial \psi}{\partial p} = \frac{2\overline{p}}{\overline{\mu Z}}$$

from the definition of pseudo-time Combining equations, we get

$$\frac{\partial \overline{p}}{\partial t} = \frac{\partial \overline{\psi}}{\partial t}\frac{\overline{\mu Z}}{2\overline{p}})$$

The denominator is calculated as follows:

$$\frac{\partial}{\partial \bar{p}}\left(\frac{\bar{p}}{\bar{z}}\right)=\bar{p}\frac{\partial}{\partial \bar{p}}\left(\frac{1}{\bar{z}}\right)+\frac{1}{\bar{z}}=-\frac{\bar{p}}{\bar{z}^2}\frac{\partial \bar{z}}{\partial \bar{p}}+\frac{1}{\bar{z}}=\frac{1}{\bar{z}}\left[1-\frac{\bar{p}}{\bar{z}}\frac{\partial \bar{z}}{\partial \bar{p}}\right]=$$

$$\frac{\bar{p}}{\bar{z}}\left(\frac{1}{\bar{p}}-\frac{1}{\bar{z}}\frac{\partial \bar{z}}{\partial \bar{p}}\right)=\frac{\bar{p}}{\bar{z}}\bar{c}_g$$

Substituting equation into collecting like terms, and simplifying, we get

$$\frac{\partial \psi}{\partial t}=\frac{\frac{2p_i}{\mu}\frac{\bar{p}}{G}\left[1-ce(pi-\bar{p})-Ei\frac{We}{G}\right]\left[-q+\frac{Eiq_{we}\left[1-\frac{Gp}{G}\right]}{\left[1-ce(pi-\bar{p})-Ei\frac{We}{G}\right]}\right]}{pi\left[1-\frac{Gp}{G}\right]ce\bar{z}+\bar{c}gzi\bar{p}\left[1-ce(pi-\bar{p})-Ei\frac{We}{G}\right]}$$

From equation, we see that

$$\left(1-\frac{Gp}{G}\right)=\frac{z_i}{p_i}\left(\frac{\bar{P}}{z}\right)\left[1-ce(pi-\bar{p})-Ei\frac{W_e}{G}\right]$$

Substituting equation simplifying, we get,

$$\frac{\partial \psi}{\partial t}=\frac{2p_i}{Zi\bar{\mu}G}\frac{\left(-q+Eiq_{we}\frac{Z_i\bar{p}}{p_i\bar{Z}}\right)}{c_e+c_g 1-c_e\left(p_i-\bar{p}\right)-Ei\frac{W_e}{G}}$$

Integrating both sides with respect to dt, we get

$$\int_{\psi_i}^{\bar{\psi}} d\psi=\frac{2p_i}{Zi\bar{\mu}G}\int_0^t\frac{\left(-q+Eiq_{we}\frac{Zi\bar{p}}{pi\bar{Z}}\right)}{c_e+c_g\left[1-c_e\left(pi-\bar{p}\right)-Ei\frac{W_e}{G}\right]}dt$$

$$\psi-\bar{\psi}=\frac{2p_i}{Z_iG}\int_0^t\frac{(q-Eq_{we})}{\mu c_{te}}dt$$

Where

$$\bar{c}_{te} = c_e + c_g\left[1 - ce\left(pi - \bar{p}\right) - Ei\frac{W_e}{G}\right]$$

$$E = \frac{V_{sc}}{V} = \frac{p}{p_{sc}}\frac{T_{sc}}{T}\frac{Z_{sc}}{Z}$$

To linearize equation, we define pseudo-time as

$$t_{ca} = \frac{(\mu c_t)i}{q}\int_0^t \frac{\left(q - Eq_{we}\right)}{\overline{\mu c}_{te}}dt$$

Now, equation is rewritten as

$$\frac{\psi_1 - \bar{\psi}}{q} = \frac{2p_i}{(\mu c_t)_i\, ZiG}t_{ca}$$

An alternate form of pseudo-time may be calculated without having to do the numerical integration. It is defined using equation as

$$t_{ca} = \frac{(\mu c_t)_i\, ZiG}{2p_i}\left[\frac{\psi_1 - \bar{\psi}}{q}\right]$$

Calculation of Reservoir Pressure and Water Influx for a Reservoir under Active Water Drive

OIL RESERVOIRS (UNDERSATURATED)

The Fetkovich water influx equation is as follows:

$$W_e = \frac{W_{ei}}{p_i}\left(pi - \bar{p}\right)\left(1 - e^{-Jp_i t / W_{ei}}\right)$$

Where

*Wei = Initial Encroachable Water = IWIP*pi*c_w*

$$J = \frac{fkh}{141.2\mu\left[\ln\frac{re}{ro} - \frac{3}{4}\right]}$$

(transfer coefficient/influx equation)

$$f = \frac{\text{encroachment angle}}{360°}$$

The oil material-balance equation (including water influx) is as follows:

$$QB_o = NB_{oi}\left(c_o\left(p_i - \bar{p}\right) + \frac{CwSw + Cf}{So}\left(p_i - \bar{p}\right)\right) + W_eB_w$$

Rearranging

$$p_i - \bar{p} = \frac{QB_o - W_eB_w}{NB_{oi}c_e}$$

Where

$$c_e = c_o + \frac{c_w s_w + c\,f}{s_o}$$

The aquifer material-balance equation is as follows:

$$p_a = p_i\left[1 - \frac{W_e}{W_{ei}}\right]$$

In order to solve for water influx and reservoir pressure, equations are discretized in time:

Aquifer Material Balance:

$$pa_{n-1} = p_i 1\left[-\frac{We_{n-1}}{W_{ei}}\right]$$

Water Influx:

$$\Delta We_n^k = \frac{Wei}{pi}\left(pa_{n-1} - \tfrac{1}{2}\bar{P}n - 1 + \bar{P}_n^k\right)\left(1 - e^{-Jpi\Delta t/Wei}\right)$$

$$We_n^k = \sum_{j=1}^{n-1} \Delta We_j + \Delta We_n^k$$

Reservoir Material Balance:

$$pi - \bar{p}_n^k = \frac{QnBo - We_n^k Bw}{NB_{oi}c_e}$$

An alternative method for solving the water influx problem, is to use a different form of the influx equation, which accounts for the aquifer material balance intrinsically, and does not assume a constant pressure difference between aqufier and reservoir. This form of the influx equation assumes a constant rate of water influx:

$$pi - \bar{p} = \frac{piWe}{Wei} + \frac{qw}{J}$$

Where qw = the rate of water influx
Discretizing equation, we get

$$pi - \bar{p}_n^k = \frac{piWe}{Wei} + \frac{We_n^k - We_{n-1}}{J\Delta t}$$

Solving equations simultaneously, we get

$$\frac{piWe_n^k}{Wei} + \frac{We_n^k - We_{n-1}}{J\Delta t} = \frac{QBo - We_n^k Bw}{NB_{oi}c_e}$$

Simplifying the above, and including the reservoir material balance, we get:

Aquifer Material Balance + Water Influx

$$We_n^k = \left(\frac{QnBo}{NB_{oi}c_e} - \frac{We_{n-1}}{J\Delta t}\right)\left(\frac{pi}{We_i} + \frac{1}{J\Delta t} + \frac{Bw}{NB_{oi}c_e}\right)^{-1}$$

Reservoir Material Balance

$$pi - \bar{p}_n^k = \frac{QnBo - We_n^k Bw}{NB_{oi}c_e}$$

GAS RESERVOIRS

The same influx equations as above also apply to gas reservoirs, except the gas material-balance equation is used, instead of oil.

$$\frac{\bar{p}}{\bar{z}} = \frac{\frac{pi}{zi}\left(1 - \frac{Gp}{G}\right)}{1 - ce\left(pi - \bar{p}\right) - Ei\frac{We}{G}}$$

The Fetkovich equations are as follows:
Aquifer Material Balance:

$$pa_{n-1} = pi\left[1 - \frac{We_{n-1}}{Wei}\right]$$

Water Influx:

$$\Delta We_n^k = \frac{Wei}{pi}\left(pa_{n-1} - \tfrac{1}{2}(\bar{p}_{n-1} + \bar{p}_n^k)\right)\left(1 - e^{-Jpi\Delta t / Wei}\right)$$

$$We_n^k = \sum_{j=1}^{n-1} \Delta We_j + \Delta We_n^k$$

Reservoir Material Balance:

$$\left(\frac{\bar{P}}{\bar{z}}\right)_n^k = \frac{\frac{pi}{zi}\left(1 - \frac{Gpn}{G}\right)}{1 - ce\left(pi - \bar{p}_n^k\right) - Ei\frac{We_n^k}{G}}$$

The modified influx equations are as follows:
Aquifer Material Balance + Water Influx

$$We_n^k = \left[pi - \bar{p}_n^k + \frac{We_{n-1}}{J\Delta t}\right]\left[\frac{pi}{Wei} + \frac{1}{J\Delta t}\right]^{-1}$$

Reservoir Material Balance

$$\left(\frac{\bar{P}}{\bar{z}}\right)_n^k = \frac{\frac{pi}{zi}\left(1 - \frac{Gpn}{G}\right)}{1 - ce\left(pi - \bar{p}_n^k\right) - Ei\frac{We_n^k}{G}}$$

AQUIFER PRODUCTIVITY INDEX

Radial Edge Water Drive

$$J = \frac{k_a h}{141.2\mu a\left(\ln\frac{ra}{re} - \frac{3}{4}\right)}$$

where:
ra = aquifer radius
re = reservoir radius
and ra/re >> 1
Mobility Ratio (M) can be calculated as follows:

$$M = \frac{ka\mu r}{\mu akr} = \frac{141.2J}{h}\left(\ln\left[\sqrt{\frac{5.615B_{wi}IWIP}{\pi\phi hre^2} + 1}\right] - \frac{3}{4}\right)\frac{\mu r}{kr}$$

Where:
re = reservoir radius (obtained from typecurve match)
ur = viscosity of reservoir fluid
kr = permeability of reservoir to reservoir fluid

For all ra/re values, the following, more rigorous derivation is applicable

$$J=\frac{kah}{141.2\mu a}\frac{1}{2\left(rad^2-1\right)}-\frac{\left(3rad^4-4rad^4\ln r_{ad}-2r_{ad}{}^2-1\right)}{4\left(rad^2-1\right)^2}$$

$$\text{where } r_{ad}=\frac{ra}{re}=\sqrt{\frac{5.615B_{wi}IWIP}{\pi\phi hr_e{}^2}+1}$$

Now,

$$M=\frac{ka\mu r}{\mu akr}=\frac{141.2J}{h}\left(\frac{1}{2\left(rad^2-1\right)}-\frac{\left(3rad^4-4rad^4\ln r_{ad}-2r_{ad}{}^2-1\right)}{4\left(rad^2-1\right)^2}\right)\frac{\mu e}{kr}$$

MULTI-PHASE (NUMERICAL)

The underlying assumption of the analytical models for production data analysis is *single phase flow* in the reservoir. In order to accommodate multiple flowing phases, the model must be able to handle changing fluid saturations and relative permeabilities.

Since these phenomena are highly non-linear, analytical solutions are very difficult to obtain and use. Thus, *numerical models* are generally used to provide solutions for the multi-phase flow problem. The numerical engine that is used in F.A.S.T. RTA is a general purpose black-oil simulator.

NUMERICAL ENGINE

Numerical models solve the nonlinear partial-differential equations (PDE's) describing fluid flow through porous media with numerical methods. Numerical methods are the process of discretizing the PDE's into algebraic equations and solving those algebraic equations to obtain the solutions.

These solutions that represent the reservoir behaviour are the values of pressure and phase saturation at discrete points in the reservoir and at discrete times. The advantages of numerical method approach are that the reservoir heterogeneity, mass transfer between phases, and forces/mechanisms responsible for flow can be taken into consideration adequately, for instance, multiphase flow, capillary and gravity forces, spatial variations of rock properties, fluid properties, and relative permeability characteristics can be represented accurately in a numerical model.

In general, analytical methods provide exact solutions to simplified problems, while numerical methods yield approximate solutions to the exact problems.

RADIAL MODEL

Radial model is a cylindrical reservoir model used for single-well studies. Cylindrical grids are used in the reservoir. The gridblock size increases lograrithmically in size outward from the well.

Small grids near the wellbore can effectively simulate the well behaviour. In current version of RTA, numerical model is a one-dimension radial model, and gas is modeled by single-phase model, oil can be modeled either by single-phase model or by multiphase model.

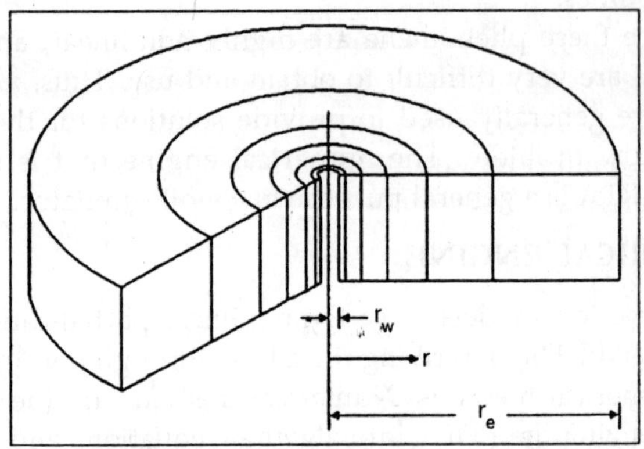

Fig. 4.16

The interface of the numerical model is almost identical to that of the analytical models, allowing maximum efficiency with minimum user input. Model parameters such as grid structure and refinement and time-step control are automatically optimized by the model.

Future additions will include:

- 2-D, Three phase, vertical/fractured well
  - Different reservoir shapes, well locations
  - Heterogeneous reservoirs
- 3-D, Three-phase
  - Layered systems
  - Gas over water, oil over water
  - Gas cap

RELATIVE PERMEABILITY

In a single-phase system, such as a dry gas reservoir or undersaturated oil reservoir, the effective permeability of the formation to the fluid varies only negligibly with time, because fluid saturations are constant.

However, when multiple phases are mobile in the reservoir, the effective permeability to a given phase will change according to the saturations of the fluids in the reservoir.

GEOMECHANICAL

Geomechanical models simulate changes in rock properties with time and pressure. In performance based reservoir analysis, a geomechanical model may be coupled with the fluid flow model (analytical or numerical). Geomechanical models are useful for (and may be necessary additions to) analyses of *overpressured* reservoirs.

The primary properties of interest are porosity, formation compressibility (pore compressibility) and permeability. The impact of these properties on the fluid flow model will be discussed the following sections.

PRESSURE DEPENDENT PERMEABILITY: MODIFIED PSEUDO-PRESSURE

In the standard pressure transient equations, permeability is usually considered to be constant. There are several situations where this may not be a valid assumption:

1. Compaction in overpressured reservoirs
2. Very low permeability reservoirs in general
3. Unconsolidated and/or fractured formations

One way to account for a variable permeability over time is to modify the definition of pseudo-pressure and pseudo-time.

The Pseudo-Steady-State equation for gas is follows:

$$\Delta p_p = p_{pi} - p_{pwf} = \frac{2qpi}{(\mu c_t Z)iGi} t_a + \frac{1.417e6^* Tq}{kh}\left(\ln\frac{re}{rwa} - \frac{3}{4}\right)$$

where

$$\Delta p_p = 2\int_{pwf}^{pi} \frac{pdp}{\mu z}$$

$$t_a = (\mu c_t)i \int_0^t \frac{dt}{\mu ct}$$

The modified PSS equation, which incorporates a variable permeability with pressure, is as follows:

$$\Delta p_p{}^* = \frac{2qpi}{(\mu c_t Z)iGi} t_a{}^* + \frac{1.417e6^* Tq}{k_i h}\left(\ln\frac{re}{rwa} - \frac{3}{4}\right)$$

where

$$\Delta p_p{}^* = \frac{2}{k_i}\int_{pwf}^{pi} \frac{k(p)pdp}{\mu z}$$

$$ta^* = \frac{(\mu c_t)i}{k_i}\int_0^t \frac{\bar{k}dt}{\mu ct}$$

Alternatively, the modified equation may be written in two parts as follows:

Depletion Part:

$$\int_p^{pi} \frac{kpdp}{\mu z} = \frac{qp_i}{Z_i G_i}\int_0^t \frac{\bar{k}dt}{\mu ct} \Rightarrow \int_p^{pi} \frac{pdp}{z} = \frac{qp_i}{Z_i G_i}\int_0^t \frac{dt}{ct}$$

Inflow Part:

$$2\int_{pwf}^{p} \frac{kpdp}{\mu z} = \frac{1.417e6 * Tq}{h}\left(\ln\frac{r_e}{r_{wa}} - \frac{3}{4}\right)$$

Equation is a material-balance equation, and thus, has no dependence on the mobility terms k and m. The mobility terms only enter into the inflow part of the PSS equation. When the PSS equation is coupled, however, the k/m term must be included in pseudo-time.

POROSITY AND FORMATION COMPRESSIBILITY

Pore volume and pore volume compressibility are related through the following.

$$c_f = \frac{1}{V_p}\frac{dVp}{dp}$$

In normally pressured reservoirs, the pore volume change with pressure is considered minimal, and thus the pore volume (formation) compressibility retains a very small, constant value. However, in overpressured reservoirs, the natural compaction process is incomplete, as a large portion of the overburden remains supported by high internal pore pressure. As this pressure is released, through fluid production, the pore space may reduce significantly.

Thus, under overpressured conditions, cf is relatively large, and may have significant variation with pressure. This becomes important to the material-balance equation, as the pore compressibility is a significant energy term in overpressured reservoirs. In normally pressured gas reservoirs, the energy of the formation is usually negligible, compared to the energy of the gas. In oil reservoirs, the formation compressibility may be significant at any pressure.

The graph below compares the energies (compressibilties) of formation and fluid (gas) over a large pressure range. In areas where they are near the same order of magnitude, the formation compressibility cannot be ignored.

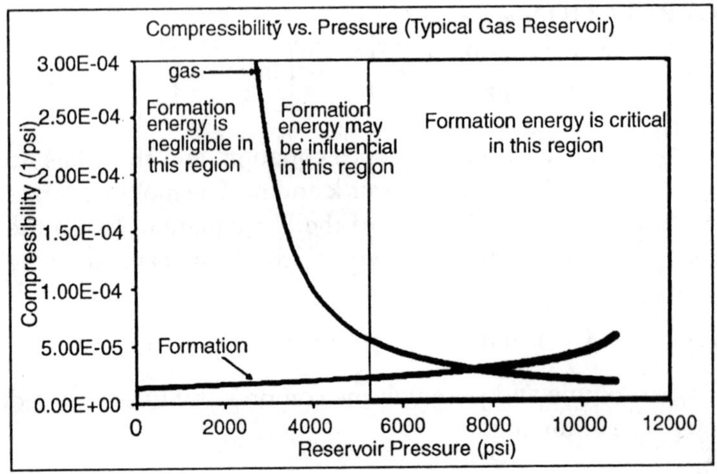

Fig. 4.17

MATERIAL BALANCE CONSIDERATIONS

In a conventional gas material balance analysis, a plot of p/z against cumulative production yields a straight line. In overpressured reservoirs, the data trend is not linear, but has some downward concavity. Thus, in the absence of a well established shut-in pressure history, the conventional p/z plot can be misleading as it often over-predicts OGIP (particularly when only early data are available).

The non-linearity of the p/z plot may be corrected by using a p/z*, which incorporates the effect of formation compressibility into the material balance. If the correct cf value is used, the p/z* plot yields a straight line. Likewise the underlying material balance of the flow model (Pseudo-steady state equation) may be compensated for formation compressibility effects.

Furthermore, a variable formation compressibility may be incorporated into the material balance within the analytical models, through the use of a modified pseudo-time function. Similar to the analytical models, a pore volume modifier may be included in the numerical models, to simulate the effect of variable formation compressibility.

CORRELATIONS

Hall

The most well-known and used correlation for formation compressibility was developed by Hall, and is a function only of porosity. The Hall correlation is based on laboratory data and is considered reasonable for normally pressured sandstones. It tends to underpredict formation compressibility under high pressure conditions.

Dobrynin

For overpressured reservoirs, there are very few published correlations. Most appear to be highly specific to area and rock-type, and thus are not useful for universal application. The Dobrynin correlation can predict a relative change in formation compressibility over a range of net overburden pressures.

It is designed for overpressured reservoirs. The correlation assumes a semi-logarithmic relationship between formation compressibility and net overburden pressure between two limiting pressures, pmin and pmax. However, it does not predict the initial formation compressibility.

$$c_f = \frac{c_f^{max}}{\log \frac{P_{max}}{P_{min}}} \log \frac{P_{max}}{P_n}$$

$$f_c = \frac{c_f}{c_{fi}} \frac{\log \frac{P_{max}}{P_n}}{\log \frac{P_{max}}{P_{n1}}}$$

where:

- c_f Pore compressibility at reservoir pressure *P*
- P_{min} Maximum net overburden pressure, psi (default = 30000psia)
- P_{max} Minimum net overburden pressure, psi (default = 150 psia)

- c_f^{max} Pore compressibility at zero net overburden pressure, 1/psi
- P_e Bulk overburden pressure $P_e = P_{grd}$ * Depth, psia
- P_n Net overburden pressure $P_n = P_e - \alpha P$, psia
- P Reservoir fluid pressure, psia
- α Biot's coefficient
- f_c Compressibility modifier

An empirical correlation for Biot Coefficient *a* is proposed in SPE 20922.

$$\alpha = 1.75\phi^{0.51}$$

The Dobrynin correlation is considered a reasonable predictor of variation in cf, provided that the following conditions are met:

- Net overburden pressure must lie within the range of pmin and pmax (usually 150 to 30,000 psi)
- Overburden pressure must be higher than the product of the initial pressure and Biot's number (initial net overburden pressure must be greater than zero)

The Dobrynin correlation may be used to generate a continuous compressibility profile using two alternatives for user inputs:

User Input Alternative 1: One cf value known:

- Initial formation compressibility (usually between 10 and 80 microcips for overpressured reservoirs)
- Overburden pressure gradient (Pgrd, psi/ft)
- Reservoir depth (ft)

Equation is then used to solve for cf at any value of p.

User Input Alternative 2: Two or more cf values known

- Overburden pressure gradient, Pgrd (default: 17 kPa/m = 0.752psi/ft)
- Reservoir depth, Depth (ft)
- Formation compressibility at reservoir pressure $P_{1,}$ $P_{2...}$, $c_{p1,}$ $c_{p2,...}$ (1/psi)

Since c_p vs. Log($_{Pn}$) is a straight line. We can do a least square regression on the data (Pi, $P_{2,}$ $P_{3...}$), ($c_{pi,}$ $c_{p2,}$ $c_{p3,...}$) for the following equation

$$c_p = a\log(P_n) + b$$

After find a and b, we can calculate C_F at any other reservoir pressure *P*. And

$$f_c = \frac{c_{p\_calc}}{c_{pi\_calc}}$$

If two data points are available, a and b can be calculated as

$$a = \frac{c_{p1} - c_{p2}}{\log\left(\frac{P_{n1}}{P_{n2}}\right)}$$

$$b = c_{p1} - a \cdot \log(P_{n1})$$

The relation between Eqs. are

$$a = -\frac{c_p^{max}}{\log \frac{P_{max}}{P_{min}}}$$

$$b = \frac{c_p^{max}}{\log \frac{P_{max}}{P_{min}}} \log P_{max}$$

PERMEABILITY

Variable permeability is an important (and sometimes critical) consideration, with regards to the fluid flow model. As a result of pore volume reduction during fluid withdrawal, the available flow area is reduced, and thus the permeability decreases with pressure.

As with formation compressibility, the variable permeability effect is greatest under overpressured conditions. The process of pore volume collapse in overpressured reservoirs is refered to as compaction. The graph below shows the effect of compaction on the permeability of several different rock samples. Variable permeability has no effect on the material balance, and thus, would not influence the interpretation of a p/z plot. However, in the flow model (PSS equation), a varying permeability could easily lead to an erroneous interpretation. As an example, consider the Flowing Material Balance plot below. The plot is generated based on production and flowing pressure data from an overpressured reservoir. For comparison, the p/z data are shown on the same graph.

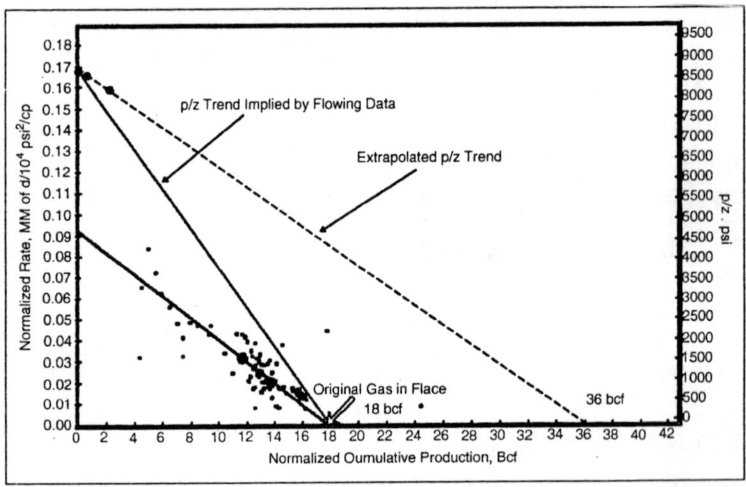

Fig. 4.18 Flowing and Conventional Material Balance Plots Compared.

It is clear from the above that the FMB extrapolates to a much lower OGIP than indicated by the p/z (note that both plots account for formation compressibility). The discrepancy is a result of the assumption of constant productivity (permeability and skin) in the FMB plot. To account for variable permeability, the analytical models may be modified by defining a new pseudo-pressure function Likewise, numerical models may be corrected for varying permeability using a permeability modifier. When analyzing flow data where pressure dependent permeability is suspected, it is critical to have access to shut-in data to compare. With flowing data alone, it is usually impossible to distinguish changing permeability (or skin) through time from volumetric depletion.

CORRELATIONS

The correlations available for relating absolute permeability to pressure are few, and usually there is no field data available. Two correlations are included here.

Yilmaz and Nur

Yilmaz and Nur present a generic correlation, suitable for the pressure variable permeability observed in extremely low permeability reservoirs.

The correlation introduces a "permeability modulus" which has a form identical to com-pressibility. The magnitude of the permeability modulus is a strong function of the formation compressibility.

$$\gamma = \frac{1}{k}\frac{dk}{dp}$$

$$\frac{k}{ki} = e^{-\gamma(pi-p)}$$

This form of pressure dependent permeability can easily be incorporated into the new definitions of pseudo-pressure and pseudo-time.

Inputs Required from User:

- Permeability Modulus (gamma); default- *formation compressibility*

Dobrynin

Dobrynin presents a more detailed correlation, designed for overpressured reservoirs, that relates the fractional change in permeability due to formation compressibility, overburden pressure and a dimensionless correlation factor, gamma. Between a certain minimum and maximum pressure range (pmin and pmax), this correlation produces a semi-log straight lin,

$$\frac{\Delta k}{k} = 2(1+\gamma)c_f{}^{max}$$

$$\left[p\min+\frac{p_o}{\log\left(\frac{p_{max}}{p_{min}}\right)}\left[\log\left(\frac{p_{max}}{p_o}\right)+0.434-\frac{p_{min}}{p_o}\left[\log\left(\frac{p_{max}}{p_{min}}\right)+0.434\right]\right]\right]$$

where

- Δk = change in permebile ty from conditions at net obp of zero
- k = permeabili ty at conditions o net obp of zero
- $p_{max} \approx 25,000 - 30,000$psi value above which k and cf remain constant
- $p_{min} \approx 150 - 300$psi value below which k and cf remain constant
- p_o = net overburden pressure $p_e - a\bar{p}$ (α = biot#~0.85 to 1)
- p_e = overburden stress
- $c_{f max}$ = cf at net overburden pressure of zero
- $0.33 < \gamma < 4$ = geometric factor, relates to pore size and disribution and compressibility

y is 0.33 for uniform pore distribution or very high pore compressibility, higher for poorly sorted sandstones with low pore compressibility.

The below equation relates the permeability at any pore pressure to that at conditions of zero net overburden pressure:

$$\frac{k_p}{k} = 1 - 2(1+\gamma)cf^{max}$$

$$\left[p_{min}+\frac{p_i-\alpha\bar{p}}{\log\left(\frac{p_{max}}{p_{min}}\right)}\left[\begin{array}{l}\log\left(\frac{p_{max}}{p_i-\alpha\bar{p}}\right)+\\ 0.434-\frac{p_{min}}{p_i-\alpha\bar{p}}\left[\log\left(\frac{p_{max}}{p_{min}}\right)+0.434\right]\end{array}\right]\right]$$

It is more suitable to have an equation that relates permeability at any pressure to permeability at initial conditions, thus we rewrite the above as follows.

$$\frac{k_p}{k_i}=\frac{\frac{k_p}{k}}{\frac{k_t}{k}}=1-2(1+\gamma)cf^{max}$$

$$\frac{\left[p_{min}+\frac{p_i-\alpha\bar{p}}{\log\left(\frac{p_{max}}{p_{min}}\right)}\left[\log\left(\frac{p_{max}}{p_i-\alpha\bar{p}}\right)+0.434-\frac{p_{min}}{p_i-\alpha\bar{p}}\left[\log\left(\frac{p_{max}}{p_{min}}\right)+0.434\right]\right]\right]}{\left[p_{min}+\frac{p_i-\alpha\bar{p}}{\log\left(\frac{p_{max}}{p_{min}}\right)}\left[\log\left(\frac{p_{max}}{p_i-\alpha pi}\right)+0.434-\frac{p_{min}}{p_i-\alpha pi}\left[\log\left(\frac{p_{max}}{p_{min}}\right)+0.434\right]\right]\right]}$$

where:

- ki = permeability at initial reservoir pressure
- Inputs Required from User
- Biot's Number (alpha); default- *0.85 to 1*
- Maximum Formation Compressibility (cfmax); *default- 30 to 80 microcips*
- Overburden Stress (pe); default- *calculated from estimated fracture gradient*

Fixed Inputs:

- Geometric Factor (gamma); calculated from correlation with cfmax
- Pmax; set at 30,000 psi
- Pmin; set at 150 psi

5

Soil P and K Levels

INTRODUCTION

It is common to hear people discuss fertilizer recommendations in terms of crop removal, plus or minus soil test buildup as if they were somehow disconnected from each other. In fact, they are two aspects of the same subject. Most of us fall into the habit of thinking that we are fertilizing plants. Except for foliar fertilizer or tree trunk injection, we do not fertilize plants... we fertilize soil.

Because of this, soil chemistry will determine how much of the applied nutrients the plants will be able to take up. If a soil is low in phosphorus (P) or potassium (K), it will tie-up or "fix" much of the applied fertilizer P and K (P_2O_5 and K_2O) into forms that are not available to the plants. This nutrient fixation is simply another way of saying that the soil is trying to build itself up in these nutrients... whether that is your intention or not.

The soil is a reservoir for the nutrients that have been applied or generated by other means over the years. The nutrients that a plant gets in any one season are likely to be ones that have been in the soil for many years.

Therefore you can think of fertilization as putting nutrients into one side of a reservoir while the plants are taking them out of another end. What happens inside of this nutrient reservoir is soil chemistry and microbiology. These processes, along with weather, determine how much access the plants have to the nutrients within the reservoir.

SOIL TEST P BUILDUP

A soil with a weak P level and a medium-to-strong P fixation capacity will convert or "fix" much of the applied fertilizer P into unavailable forms and leave little for the crop. In many medium P soils as much as 65% of the applied fertilizer P can be fixed. In extremely poor soils, as much as 90% of applied P may be fixed.

In these cases, applying only "crop removal" is inviting P deficiency. In such soils the P fixation capacity must be overcome by higher fertilizer P rates in order to have enough P left over for the crop.

In other words, soils with less than optimum P levels must have some soil buildup P, to insure that there is enough for the immediate needs of the crop or other plants.

The difficulty lies in determining, with some accuracy, the minimum amount of buildup P required to overcome a particular soils fixation capacity during the season. Soil test buildup application rates are typically discussed in terms of the ratio of applied P_2O_5 to soil test P buildup amounts. Often you will hear that most soils require about 9 or 10 lb. of P_2O_5, in excess of crop needs, in order to increase the soil test P by 1 lb/a. In ppm this would be 4.5 to 5 lb of P_2O_5 to raise the soil P 1 ppm.

Agronomists normally refer to this as the buildup ratio. A buildup ratio of 9 lb. P_2O_5 to build up the soil test by 1 lb/a soil test P would be simplified to 9:1 or sometimes just the number 9.

In ppm this same ratio would be stated as 4.5:1 or just 4.5. If you have heard agronomists quote this 9:1 or 4.5:1 ratio many times you might think that this is all that you need to know... it isn't that simple.

Research at the University of Kentucky reported in 2002 illustrates how soil test P buildup occurs. In this study, they looked at 16 different soils with initial soil test P levels ranging from 6 to 240 lb P/acre (3 to 120 ppm P/a) using Mehlich 3 (M3) extraction. They applied 6 different rates of phosphorus ranging from 38 to 228 lb P_2O_5/acre. Their results nicely fit a mathematical relationship used to develop the data in Table.

These results clearly show that the initial soil P test is a major factor in determining how much P_2O_5 is required to buildup the soil test P.

Initial Soil P (lb/a)	Buildup Ratio lb P_2O_5 per lb Soil P Increase	Initial Soil P (lb/a)	Buildup Ratio lb P_2O_5 per lb Soil P Increase
5	30.1	65	5.1
10	18.7	70	4.9
15	14.1	75	4.7
20	11.6	80	4.5
25	9.9	85	4.3
30	8.8	90	4.1
35	7.9	95	4.0
40	7.2	100	3.8
45	6.6	105	3.7
50	6.2	110	3.6
55	5.8	115	3.5
60	5.4	120	3.4

While other soils in different locations may have somewhat different results, the general trend will probably be similar. Notice that when the soil P test is very low, the soil P buildup ratio is very high.

However, as the initial soil P level increases, the buildup ratio gets lower. You can see by this data that the most common soil P levels found in farm fields that need some buildup have a buildup ratio of approximately 9:1 or maybe a little lower.

Work by Kamprath in 1964 illustrates that higher individual rates of application may also affect the efficiency of buildup. While we would expect higher annual rates of P_2O_5 to result in a higher soil test, this work indicates that the higher annual application rates may also be more efficient (a lower buildup ratio) at raising soil test P levels over time. In other words, with higher annual rates, it may take less total fertilizer P to reach the ultimate soil test goal.

Table. Soil P Buildup Speeds varies with rate of P205

Applied P205 (7 yr.)		Soil Test P (lb./acre)			
Annual	Total	Initial	Final	Increase	P205/P
44	306	36	50	14	21.9
87	612	36	76	40	15.3
131	918	36	116	80	11.5

SOIL TEST P DRAWDOWN

Soil test P drawdown works similar to buildup, but in reverse. Also like buildup, there is no single number or calculation that will work in all situations. Another University of Kentucky study looked at the effect of growing alfalfa for several years without fertilizer on soil test P (STP) drawdown.

In this study, 16 sites on a single farm received no fertilizer between May 26, 1998 and Feb 23, 2000. During this time the existing alfalfa was harvested in the normal manner and analysed for P removal.

The researchers reported that average P_2O_5 removal was 13.9 lb P_2O_5/T. An average of 65 lb P/a (150 lb P_2O_5) was removed from the plots and the STP was reduced an average of 19 lb/a for an average P_2O_5:P drawdown ratio of 11.8 to 1.

The range of total P_2O_5 removal was from 128 to 169 lb/acre and the average STP drawdown ranged from 4 to 44 lb P/acre for a range in drawdown ratios of from 3.2:1 to 37.8:1.

This data did not fit into a simple equation which could generate a simplified table or rule-of-thumb for us to use, so producers are advised to simply monitor their soils with routine soil tests in order to properly manage their soil fertility.

Table. Effect of P Removal by Alfalfa (Dry Matter Basis) on Change in Soil Test P

		1998			1999		2000				
	STP	P	STP	STP	P	STP	STP	Total P	Total P_2O_5		Drawdown
	(lb/A)	Removal	(lbs/A)	(lb/A)	Removal	(lb/A)	lb/A	Removal	Removal	Total STP	Ratio
Plot	5/26/98	lb/A	11/05/98	04/08/99	lbs/A	10/25/99	2/23/00	(98 & 99)	98 & 99)	Reduction	(P_2O_5/STP chg)
9	95	34	51	47	27	46	51	61	140	44	3.2
5	77	35	60	49	27	39	48	62	142	29	4.9
8	88	37	89	70	34	41	56	71	163	32	5.1
1	57	34	45	44	27	26	34	61	140	23	6.1
10	62	37	50	44	29	36	38	66	151	24	6.3
7	76	37	64	60	29	45	55	66	151	21	7.2
14	93	38	70	79	27	72	73	65	149	20	7.4
12	85	39	88	63	32	50	65	71	163	20	8.1
11	53	35	43	41	31	29	35	66	151	18	8.4
3	54	37	49	40	32	29	39	69	158	15	10.5
6	55	34	37	43	27	32	43	61	140	12	11.6
16	92	38	76	82	34	84	78	72	165	14	11.8
4	37	35	36	26	25	23	27	60	137	10	13.7
2	32	32	20	25	24	19	25	56	128	7	18.3
15	83	39	86	79	35	66	77	74	169	6	28.2
13	77	34	79	80	32	62	73	66	151	4	37.8
Low	32	32	20	25	24	19	25	56	128	4	3.2
High	95	39	89	82	35	84	78	74	169	44	37.8
Avg	70	36	59	55	30	44	51	65	150	19	11.8

Note: The data on each of the Low and High lines represent the respective values from all of the 16 plots should not be viewed as representing a single plot in any case. Wells and Dollarhide, 2002. Mehlich-3 extraction.

SOIL TEST P GOALS

Setting soil test P goals can be more complicated that it may seem at first. Many factors should be considered. For example, different crops normally require different soil P levels for best yields. For example, wheat requires a much higher soil P test than corn, while soybeans seem to grow well at a much lower soil P test than corn. At first glance, you could conclude that each crop should have a different soil P goal.

However, a field can have only one soil P level, so which one do you choose if the three crops occur in rotation? Some producers of acid-requiring crops like potatoes have another problem. An acid soil pH causes much more soil P tie-up than a pH between 6.0 and 7.0. In these situations, a higher than "normal" soil P level is often beneficial.

Table. Soil P "Good" Range

Target pH >6.0						
CEC	5	10	15	20	25	30
Soil P Range (lb./a)						
	80-120	69-109	62-99	55-93	49-88	43-84
Soil P Range (ppm)						
	40-60	34-55	31-50	27-47	25-44	22-42
Target pH<6.0						
CEC	5	10	15	20	25	30
Soil P Range (lb./a)						
	140-210	120-180	109-169	95-153	89-148	83-144
Soil P Range (ppm)						
	70-105	110-140	55-134	47-126	45-74	41-172

Note: Recommendations are based on the center of a given range. Mehlich-3 extraction

Over the years some researchers have noticed that the "ideal" soil P test appears to decrease as the soil CEC increases. While this does not appear to be a large effect, and it is not widely used, Spectrum Analytic has incorporated it into our recommendation program. The information describes our method of defining a desirable soil P test for most crop rotations and situations.

SOIL TEST K BUILDUP

While the chemistry that controls soil P and K is much different, the ultimate effects of building or drawing down the soil test of either have similar patterns. With soil test K, most agronomists use a K buildup rate of 3 or 4 lb./acre K_2O: 1 lb./acre soil K (1.5 or 2 lb K_2O:1 ppm soil K).

However, research indicates that, like P it depends somewhat on what the initial soil test is, plus other factors. Research at the University of Kentucky (Table 5) looked at this question in 1990 and found the following buildup ratios for one soil.

Table. Amounts of Excess Addition Over Removal Required To Change Soil Test K (Lb K_2O/Lb Soil Test K) At A Given Soil Test.

Initial Soil Test K (lb K/acre)	K_2O Required (lb K_2O/lb K)
100	6.4
150	5.4
200	4.7
250	4.2
300	3.8

The data in Table is taken from a single soil type; therefore variables such as the different types and amounts clay are not much of a factor. The same experiment performed in other soil types would likely show somewhat different results, but would likely have similar patterns. It would be reasonable to assume that soil test K drawdown would have patterns and ratios similar to those of buildup.

ould likely show somewhat different results, but would likely have similar patterns. It would be reasonable to assume that soil test K drawdown would have patterns and ratios similar to those of buildup. A South African study looked at the effects of the soil physical properties on soil test K buildup in 51 different soils. Unfortunately, they did not include the initial soil K test level in this study. However, their results do shed light on some other aspects of soil K buildup. They found that on average it required 3.07 lb K_2O to raise the soil test 1 lb of K.

However, the range of needed K_2O was from a low of 1.72 to a high of 10.51. Of the various soil physical factors evaluated they found the most significant factors causing a higher buildup ratio were the overall soil CEC, the CEC of only the clay fraction of the soil, a mathematical value called the K desorption index, and a mathematical factor relating the density of the soil to the CEC of the soil.

While none of these factors was dominant, the relationships showed that higher CEC soils, and those dominated by certain types of clay required higher rates of K_2O to increase the soil a given amount.

The researchers found, as had previous work, that mica and vermiculite type clay greatly increased the amount of fertilizer K_2O required to increase the soil test K. Next down on the scale were other 2:1 type clays and highly weathered 1:1 type clays had the least effects.

While these different types of clay can occur nearly everywhere, Northern soils tend to be dominated with the more complex, K-fixing types and Southern soils tend to have more of the highly weathered, less complex 1:1 type clay which is less K-fixing. Examples of the variability in the K fixation capability of different soil types is illustrated by some results published from work at Purdue and Michigan State University.

Table. Soil Type vs. Applied K_2O Fixation

Soil Type	% Fixed
Zanesville Silt Loam	7 to 37
Vigo Silt Loam	10 to 20
Crosby Silt Loam	11 to 33
Chalmers Silty Clay Loam	16 to 36
Brookston Silty Clay Loam	58 to 68

Chalmers Silty Clay Loam	16 to 36
Brookston Silty Clay Loam	58 to 68

Note: Purdue Univ., Initial soil K levels not reported. Application Rate = 400 lbs/A K20, Incubation period = 7 mo.

Table. Soil Type vs. Applied K_2O Fixation

Soil Type	% Fixed
Granby Sandy Loam	22
Genesee fine-Loamy sand	92

Note: Michigan State Univ., Initial soil K level, K_2O rate, and incubation period not reported for MI data.

The wide range in the K-fixation capacity of soils within a single State illustrates why no single buildup factor will be correct in all situations. Unfortunately, this data did not include some of the information that might make the data even more valuable, such as initial soil K test or in Michigan data, the rates of K_2O application.

However, from this you can appreciate that if 30% of the applied K_2O is fixed into an unavailable form and crop removal must be subtracted from the remaining 70%, the amount left-over to increase the available soil test K may not be very large. If your soil has a K-fixation capacity in the 60% to 90% range, you can imagine the difficulty in producing good crops or increasing the available soil K.

SOIL TEST K DRAWDOWN

The University of Kentucky study that looked at soil test P drawdown also looked at soil test K (STK) drawdown. As with P, they looked at the effect of growing alfalfa for several years without fertilizer. A total of 10.8 tons/a was harvested over the two years with a significant droughty period in each year. An average of 51.8 lb K_2O/ton was removed in the harvest. As table shows, the ratio of K_2O removed to STK drawdown averaged 3.7 (3.7:1), with a range of from 1.4 to 8.1. These ratios are similar to those found when building up soil test K. Unfortunately the researchers did not report the CEC of the soils in the study.

	1998				1999		2000				
	STK (lbs/ A)	K Remova[2]	STK (lbs/A)	STK (lbs/A)	K Removal[1]	STK (lbs/A)	STK (lbs/A)	Total K	Total K_2O	Total STK	Drawdown Ratio
Plot	5-26-98[1]	lbs/A	11/05/98[3]	04-08-99[4]	lbs/A	10/25/99	2/23/00	Remo-val	Remo-val	Reduc-tion	(K_2O/STK chg)
9	643	262	286	280	163	452	285	425	510	358	1.4
8	633	294	442	340	232	381	341	526	631	292	2.2
5	475	253	278	227	142	225	279	395	474	196	2.4
1	417	234	302	222	160	181	231	394	473	186	2.5
14	503	304	325	330	203	299	298	507	608	205	3.0
7	547	298	381	328	178	304	362	476	571	185	3.1
14	459	251	289	235	204	448	295	455	546	164	3.3
10	467	300	317	284	178	264	303	478	574	164	3.5
13	514	281	404	329	215	621	356	496	595	158	3.8
15	498	316	410	349	265	331	317	581	697	181	3.9
16	464	287	279	295	204	346	309	491	589	155	3.8
6	395	230	263	229	149	257	282	379	455	113	4.0
12	497	290	416	282	194	347	369	484	581	128	4.5
3	495	269	335	279	187	280	387	456	547	108	5.1
2	339	203	240	205	219	249	242	422	506	97	5.2
4	445	274	404	281	224	256	371	498	598	74	8.1
Low	339	203	240	205	142	181	231	379	455	74	1.4
High	643	316	442	349	265	621	387	581	697	358	8.1
Avg	487	272	336	281	195	328	314	466	560	173	3.7

Missing the CEC meant that we could not evaluate any relationship between soil CEC or K saturation and the soil test drawdown. However, there did seem to be a relationship between the initial soil test K in simple lb/a and the drawdown ratio. This relationship indicated that the drawdown ratio would tend to be higher with a lower initial soil test K.

This would seem to be logical since as a soil test approaches its "natural" base level, it becomes increasingly difficult to decrease the soil test K further. Typically, the "natural" base or minimum soil test K level will be lower on a low CEC soil and higher as CEC increases.

SOIL K GOALS

The following data lists some soil CEC and soil K levels used by Spectrum Analytic. While agronomists may differ, most would consider these to be adequate-to-high soil K goals.

Using the previous information, you can estimate the amount of K_2O required to achieve these soil K goals. Don't forget that soil test buildup amounts occur only after accounting for crop removal.

Table. Desirable Soil K Test Levels

		Soil K(lb/a)	Soil K (ppm)	K Saturation (% sat.)
CEC	5	180-270	90-135	4.6-6.9
	10	240-380	120-190	3.1-4.9
	15	300-460	150-230	2.6-3.9
	20	340-530	170-265	2.2-3.4
	25	380-570	190-285	1.9-2.9
	30	400-615	200-308	1.7-2.6

Note:

Mehlich-3 extraction

TYPES OF COMPACTION

There are four types of compaction effort on soil or asphalt:

1. Vibration
2. Impact

3. Kneading
4. Pressure

These different types of effort are found in the two principle types of compaction force: static and vibratory. Static force is simply the deadweight of the machine, applying downward force on the soil surface, compressing the soil particles. The only way to change the effective compaction force is by adding or subtracting the weight of the machine.

Static compaction is confined to upper soil layers and is limited to any appreciable depth. Kneading and pressure are two examples of static compaction. Vibratory force uses a mechanism, usually engine-driven, to create a downward force in addition to the machine's static weight. The vibrating mechanism is usually a rotating eccentric weight or piston/ spring combination (in rammers). The compactors deliver a rapid sequence of blows (impacts) to the surface, thereby affecting the top layers as well as deeper layers. Vibration moves through the material, setting particles in motion and moving them closer together for the highest density possible. Based on the materials being compacted, a certain amount of force must be used to overcome the cohesive nature of particular particles.

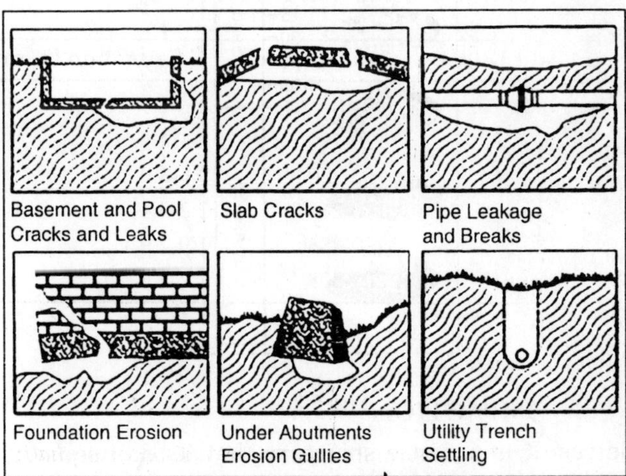

Fig.5.1 Results of Poor Compaction

SOIL TYPES AND CONDITIONS

Every soil type behaves differently with respect to maximum density and optimum moisture. Therefore, each soil type has its own unique requirements and controls both in the field and for testing purposes. Soil types are commonly classified by grain size, determined by passing the soil through a series of sieves to screen or separate the different grain sizes. Soil classification is categorized into 15 groups, a system set up by AASHTO. Soils found in nature are almost always a combination of soil types. A well-graded soil consists of a wide range of particle sizes with the smaller particles filling voids between larger particles.

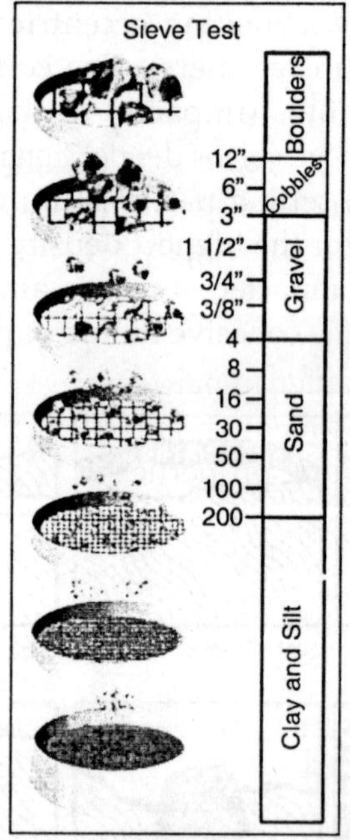

Fig. 5.2

The result is a dense structure that lends itself well to compaction. A soil's makeup determines the best compaction method to use.

The are three basic soil groups:

1. Cohesive
2. Granular
3. Organic (this soil is not suitable for compaction and will not be discussed here)

COHESIVE SOILS

Cohesive soils have the smallest particles. Clay has a particle size range of.00004" to.002". Silt ranges from.0002" to.003". Clay is used in embankment fills and retaining pond beds.

Characteristics

Cohesive soils are dense and tightly bound together by molecular attraction. They are plastic when wet and can be molded, but become very hard when dry.

Proper water content, evenly distributed, is critical for proper compaction. Cohesive soils usually require a force such as impact or pressure. Silt has a noticeably lower cohesion than clay. However, silt is still heavily reliant on water content.

GRANULAR SOILS

Granular soils range in particle size from.003" to.08" (sand) and.08" to 1.0" (fine to medium gravel). Granular soils are known for their water-draining properties.

Characteristics

Table. Guide to Soil Types

What to look for	Appearance/Feel	Water Movement	When Moist	When Dry
Granular soils, fine sands and silts	Coarse grains can be seen. Feels gritty when rubbed between fingers	When water and soil are shaken in palm of hand, they mix. When shaking is stopped theyseparate	Very little or no plasticity	Little or no cohesive strength when dry. Soil sample will crumble easily.
Cohesive soils, mixes and clays	Grains cannot be seen by naked eye. Feels smooth and greasy when rubbed between fingers	When water and soil are shaken in palm of hand, they will not mix	Plastic and sticky. Can be rolled	Has high strength when dry. Crumbles with difficulty. Slow saturation in water.

Sand and gravel obtain maximum density in either a fully dry or saturated state. Testing curves are relatively flat so density can be obtained regardless of water content. The tables that follow give a basic indication of soils used in particular construction applications.

Table. Materials

		Vibrating Sheepsfoot Rammer	Static Sheepsfoot Grid Roller Scraper	Vibrating Plate Compactor Vibrating Roller Vibrating Sheepsfoot	Scraper Rubber-tired Roller Loader Grid Roller
	Lift Thickness	**Impact**	**Pressure (with kneading)**	**Vibration**	**Kneading (with pressure)**
Gravel	12+	Poor	No	Good	Very Good
Sand	10+/-	Poor	No	Excellent	Good
Silt	6+/-	Good	Good	Poor	Excellent
Clay	6+/-	Excellent	Very Good	No	Good

Table. Fill Materials

	Permeability	Foundation Support	Pavement Sub grade	Expansive	Compaction Difficulty
Gravel	Very High	Excellent	Excellent	No	Very Easy
Sand	Medium	Good	Good	No	Easy
Silt	Medium Low	Poor	Poor	Some	Some
Clay	None+	Moderate	Poor	Difficult	Very Difficult
Organic	Low	Very Poor	Not Acceptable	Some	Very Difficult

EFFECT OF MOISTURE

The response of soil to moisture is very important, as the soil must carry the load year-round. Rain, for example, may transform soil into a plastic state or even into a liquid. In this state, soil has very little or no load-bearing ability.

Moisture vs. Soil Density

Moisture content of the soil is vital to proper compaction. Moisture acts as a lubricant within soil, sliding the particles together. Too little moisture means inadequate compaction - the particles cannot move past each other to achieve density. Too much moisture leaves water-filled voids and subsequently weakens the load-bearing ability. The highest density for most soils is at a certain water content for a given compaction effort. The drier the soil, the more resistant it is to compaction. In a water-saturated state the voids between particles are partially filled with water, creating an apparent cohesion that binds them together. This cohesion increases as the particle size decreases (as in clay-type soils).

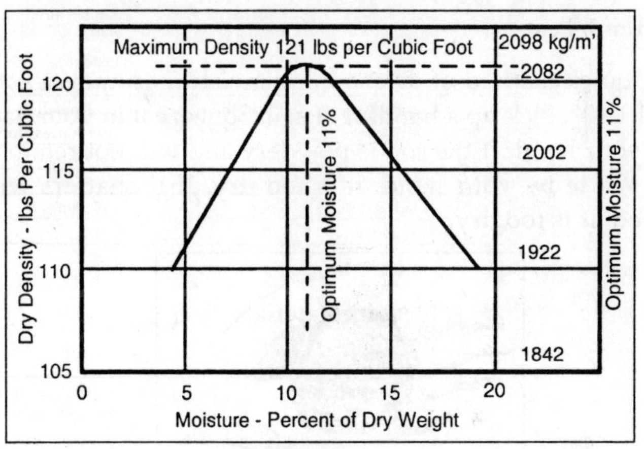

Fig. 5.3

Soil Density Tests

To determine if proper soil compaction is achieved for any specific construction application, several methods were developed. The most prominent by far is soil density.

Why Test

Soil testing accomplishes the following:

- Measures density of soil for comparing the degree of compaction vs. specs
- Measures the effect of moisture on soil density vs. specs
- Provides a moisture density curve identifying optimum moisture

TYPES OF TESTS

Tests to determine optimum moisture content are done in the laboratory. The most common is the Proctor Test, or Modified Proctor Test. A particular soil needs to have an ideal (or optimum) amount of moisture to achieve maximum density. This is important not only for durability, but will save money because less compaction effort is needed to achieve the desired results.

The Hand Test

A quick method of determining moisture is known as the "Hand Test". Pick up a handful of soil. Squeeze it in your hand. Open your hand. If the soil is powdery and will not retain the shape made by your hand, it is too dry. If it shatters when dropped, it is too dry.

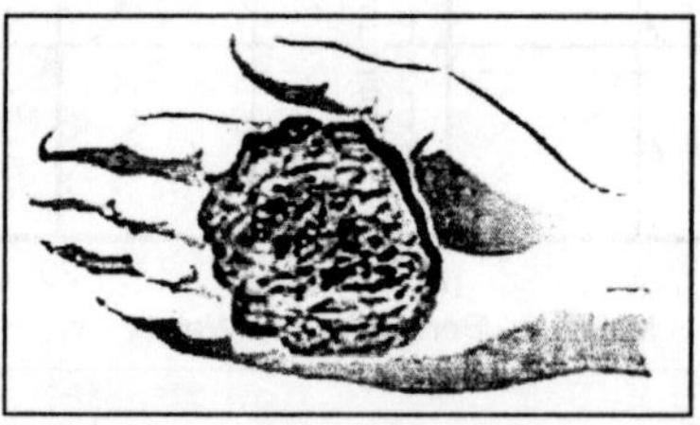

Fig. 5.4

If the soil is moldable and breaks into only a couple of pieces when dropped, it has the right amount of moisture for proper compaction. If the soil is plastic in your hand, leaves traces of moisture on your fingers and stays in one piece when dropped, it has too much moisture for compaction.

Proctor Test (ASTM D1557-91)

The Proctor, or Modified Proctor Test, determines the maximum density of a soil needed for a specific job site. Secondly, it tests the effects of moisture on soil density. These values are determined before any compaction takes place to develop the compaction specifications. Modified Proctor values are higher because they take into account higher densities needed foe certain typed of construction projects. Test methods are similar for both tests. A small soil sample is taken from the jobsite. A standard weight is dropped several times on the soil. The material weighed and then oven dried for 12 hours in order to evaluate water content

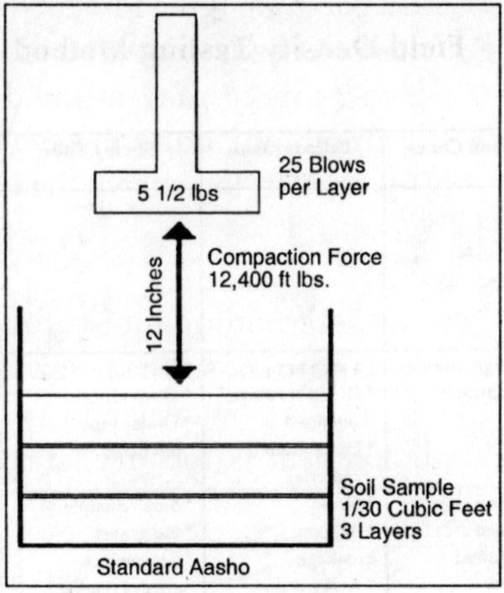

Fig. 5.5

This is similar to the Proctor Test except a hammer is used to compact material for greater impact, The test is normally preferred in testing materials for higher shearing strength.

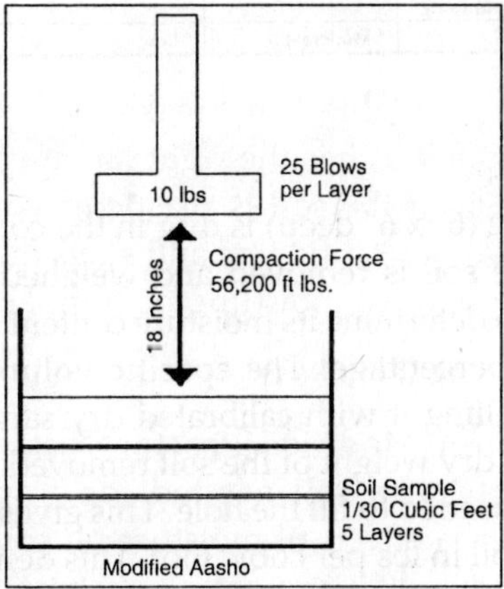

Fig. 5.6

Field Tests

It is important to know and control the soil density during compaction. Following are common field tests to determine on the spot if compaction densities are being reached.

Field Density Testing Method

	Sand Cone	Balloon Dens meter	Shelby Tube	Nuclear Gauge
Advantages	* Large sample * Accurate	* Large sample * Direct reading obtained * Open graded material	* Fast * Deep sample * Under pipe haunches	* Fast * Easy to redo * More tests (statistical reliability)
Disadvantages	* Many steps * Large area required * Slow * Halt Equipment * Tempting to accept flukes	* Slow * Balloon breakage * Awkward	* Small Sample * No gravel * Sample not always retained	* No sample * Radiation * Moisture suspect * Encourages amateurs
Errors	* Void under plate * Sand bulking * Sand compacted * Soil pumping	* Surface not level * Soil pumping * Void under plate	* Overdrive * Rocks in path * Plastic soil	* Miscalibrated * Rocks in path * Surface prep required * Backscatter
Cost	* Low	* Moderate	* Low	* High

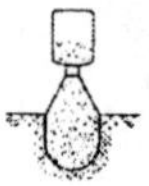

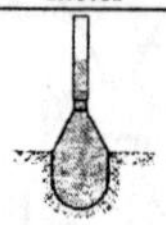

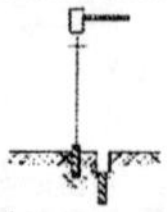

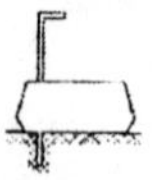

Sand Cone Test

A small hole (6" x 6" deep) is dug in the compacted material to be tested. The soil is removed and weighed, then dried and weighed again to determine its moisture content. A soil's moisture is figured as a percentage. The specific volume of the hole is determined by filling it with calibrated dry sand from a jar and cone device. The dry weight of the soil removed is divided by the volume of sand needed to fill the hole. This gives us the density of the compacted soil in lbs per cubic foot. This density is compared to the maximum Proctor density obtained earlier, which gives us the relative density of the soil that was just compacted.

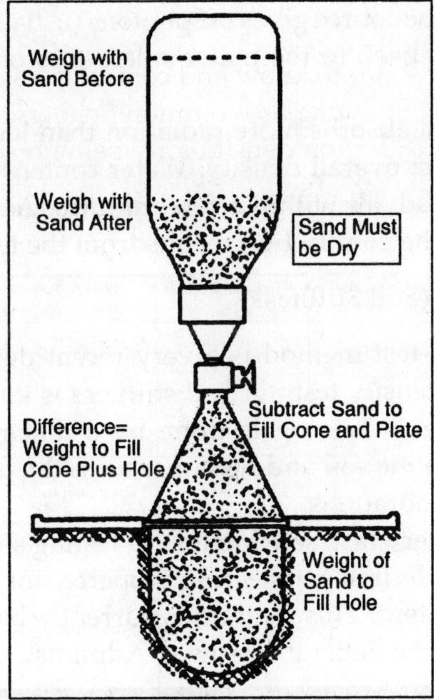

Fig. 5.7

Nuclear Density

Nuclear Density meters are a quick and fairly accurate way of determining density and moisture content. The meter uses a radioactive isotope source (Cesium 137) at the soil surface (backscatter) or from a probe placed into the soil (direct transmission).

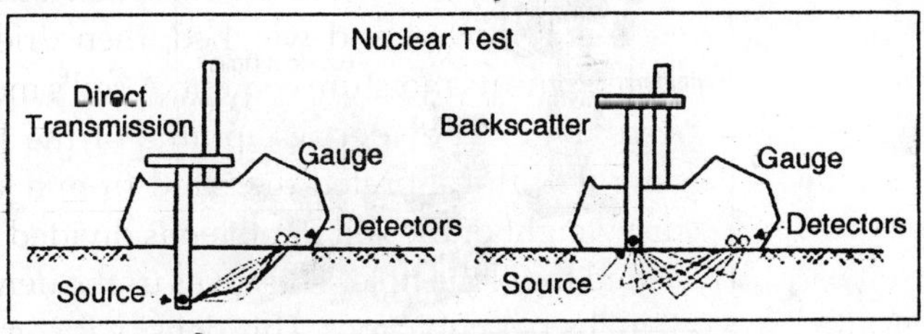

Fig. 5.8

The isotope source gives off photons (usually Gamma rays) which radiate back to the mater's detectors on the bottom of the unit.

Dense soil absorbs more radiation than loose soil and the readings reflect overall density. Water content (ASTM D3017) can also be read, all within a few minutes. A relative Proctor density with the compaction results from the test.

Soil Modulus (Soil Stiffness)

This field-test method is a very recent development that replaces soil density testing. Soil stiffness is the ratio of force-to-displacement. Testing is done by a machine that sends vibrations into the soil and then measures the deflection of the soil from the vibrations.

This is a very fast, safe method of testing soil stiffness. Soil stiffness is the desired engineering property, not just dry density and water content. This method is currently being researched and tested by the Federal Highway Administration.

COMPACTION EQUIPMENT

APPLICATIONS

The desired level of compaction is best achieved by matching the soil type with its proper compaction method. Other factors must be considered as well, such as compaction specs and job site conditions.

Fig. 5.9

Cohesive soils: Clay is cohesive, its particles stick together. Therefore, a machine with a high impact force is required to ram the soil and force the air out, arranging the particles. A rammer is the best choice, or a pad-foot vibratory roller if higher production is needed. The particles must be sheared to compact.

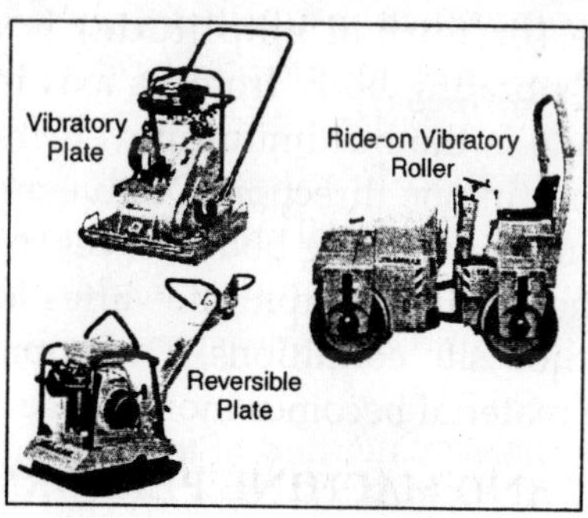

Fig. 5.10

Granular soils: Since granular soils are not cohesive and the particles require a shaking or vibratory action to move them, vibratory plates (forward travel) are the best choice. Reversible plates and smooth drum vibratory rollers are appropriate for production work. Granular soil particles respond to different frequencies (vibrations) depending on particle size. The smaller the particle, the higher the frequencies and higher compaction forces. Normally, soils are mixtures of clay and granular materials, making the selection of compaction equipment more difficult. It is a good idea to choose the machine appropriate for the larger percentage of the mixture. Equipment testing may be required to match the best machine to the job. Asphalt is considered granular due to its base of mixed aggregate sizes (crushed stone, gravel, sand and fines) mixed with bitumen binder (asphalt cement). Consequently, asphalt must be compacted with pressure (static) or vibration.

Compaction Machine Characteristics

Two factors are important in determining the type of force a compaction machine produces: frequency and amplitude.

Frequency is the speed at which an eccentric shaft rotates or the machine jumps. Each compaction frequency machine is designed to operate at an optimum frequency to supply the maximum force. Frequency is usually given in terms of vibration per minute (vpm).

Amplitude (or normal amplitude) is the maximum movement of a vibrating body from its axis in one direction. Double amplitude is the maximum movement of a vibrating body from its axis in one direction. Double amplitude ins the maximum distance a vibrating body moves in both directions from its axis. The apparent amplitude varies for each machine under different job site conditions. The apparent amplitude increases as the material becomes more dense and compacted.

LIFT HEIGHT AND MACHINE PERFORMANCE

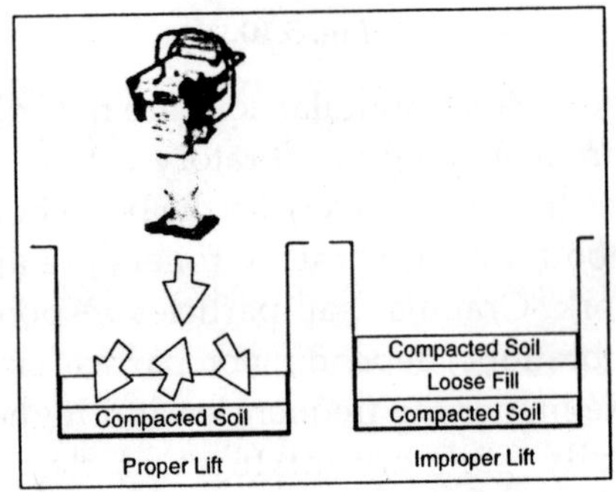

Fig. 5.11

Lift height (depth of the soil layer is an important factor that effectsmachine performance and compaction cost. Vibratory and rammer-type equipment compact soil in the same direction: from top to bottom and bottom to top. As the machine hits the soil, the impact travels to the hard surface below and then returns upward. This sets all particles in motion and compaction takes place. As the soil becomes compacted, the impact has a shorter distance to travel. More force returns to the machine, making it lift off the ground higher in its stroke cycle. If the lift is too deep, the machine will take longer to compact the soil and a layer within the lift will not be compacted. Soil can also be over-compacted if the compactor makes too many passes (a pass is the machine going across a lift in one direction). Over-compaction is like constantly hitting concrete with a sledgehammer. Cracks will eventually appear, reducing density. This is a waste of man-hours and adds unnecessary wear to the machine.

COMPACTION SPECIFICATIONS

A word about meeting job site specifications. Generally, compaction performance parameters are given on a construction project in one of two ways: - Method Specification - detailed instructions specify machine type, lift depths, number of passes, machine speed and moisture content. A "recipe" is given as part of the job spec to accomplish the compaction needed.

This method is outdated, as machine technology has far outpaced common method specification requirements. - End-result Specification - engineers indicate final compaction requirements, thus giving the contractor much more flexibility in determining the best, most economical method of meeting the required specs. Fortunately, this is the trend, allowing the contractor to take advantage of the latest technology available.

EQUIPMENT TYPES

Rammers

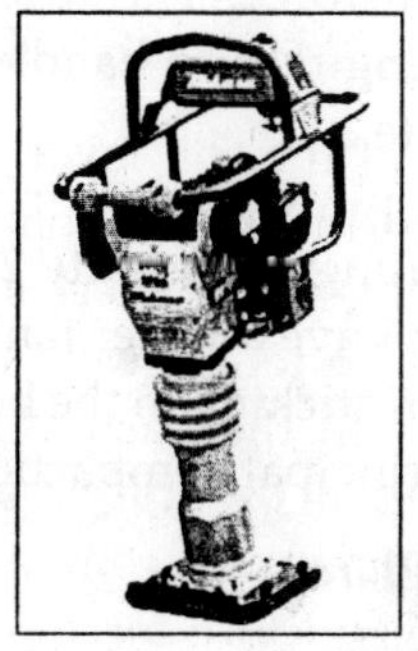

Fig. 5.12

Rammers deliver a high impact force (high amplitude) making them an excellent choice for cohesive and semi-cohesive soils. Frequency range is 500 to 750 blows per minute. Rammers get compaction force from a small gasoline or diesel engine powering a large piston set with two sets of springs.

The rammer is inclined at a forward angle to allow forward travel as the machine jumps. Rammers cover three types of compaction: impact, vibration and kneading.

Vibratory Plates

Fig. 5.13

Vibratory plates are low amplitude and high frequency, designed to compact granular soils and asphalt. Gasoline or diesel engines drive one or two eccentric weights at a high speed to develop compaction force. The resulting vibrations cause forward motion. The engine and handle are vibration-isolated from the vibrating plate.

The heavier the plate, the more compaction force it generates. Frequency range is usually 2500 vpm to 6000 vpm. Plates used for asphalt have a water tank and sprinkler system to prevent asphalt from sticking to the bottom of the base plate. Vibration is the one principal compaction effect.

Reversible Vibratory Plates

In addition to some of the standard vibratory plate features, reversible plates have two eccentric weights that allow smooth transition for forward or reverse travel, plus increased compaction force as the result of dual weights. Due to their weight and force, reversible plates are ideal for semi-cohesive soils.

A reversible is possible the best compaction buy dollar for dollar. Unlike standard plates, the reversible forward travel may be stopped and the machine will maintain its force for "spot" compaction.

ROLLERS

Fig. 5.14

Rollers are available in several categories: walk-behind and ride-on, which are available as smooth drum, padded drum, and rubber-tired models; and are further divided into static and vibratory sub-categories.

Walk-behind

Smooth A popular design for many years, smooth-drum machines are ideal for both soil and asphalt. Dual steel drums are mounted on a rigid frame and powered by gasoline or diesel engines. Steering is done by manually the machine handle. Frequency is around 4000 vpm and amplitudes range from.018 to.020.

Vibration is provided by eccentric shafts placed in the drums or mounted on the frame. Padded rollers are also known as trench rollers due to their effective use in trenches and excavations. These machines feature hydraulic or hydrostatic steering and operation. Powered by diesel engines, trench rollers are built to withstand the rigors of confined compaction.

Trench rollers are either skid-steer or equipped with articulated steering. Operation can be by manual or remote control. Large eccentric units provide high impact force and high amplitude (for rollers) that are appropriate for cohesive soils. The drum pads provide a kneading action on soil. Use these machines for high productivity.

RIDE-ON

Configured as static-wheel rollers, ride-ons are used primarily for asphalt surface sealing and finishing work in the larger (8 to 15 ton) range. Small ride-on units are used for patch jobs with thin lifts. The trend is towards vibratory rollers. Tandem vibratory rollers are usually found with drum widths of 30" up to 110", with the most common being 48".

Suitable for soil, sub-base and asphalt compaction, tandem rollers use the dynamic force of eccentric vibrator assemblies for high production work. single-drum machines feature a single vibrating drum with pneumatic drive wheels. The drum is available as smooth for sub-base or rock fill, or padded for soil compaction.

Additionally, a ride-on version of the pad foot trench roller is available for very high productivity in confined areas, with either manual or remote control operation. Rubber-tire These rollers are equipped with 7 to 11 pneumatic tires with the front and rear tires overlapping.

A static roller by nature, compaction force is altered by the addition or removal of weight added as ballast in the form of water or sand.

Weight ranges vary from 10 to 35 tons. The compaction effort is pressure and kneading, primarily with asphalt finish rolling. Tire pressures on some machines can be decreased while rolling to adjust ground contact pressure for different job conditions.

Table. Equipment Applications

	Granular Soils	Sand and Clay	Cohesive Clay	Asphalt
Rammers	Not Recommended	Testing Recommended	Best Application	Not Recommended
Vibratory Plates	Best Application	Testing Recommended	Not Recommended	Best Application
Reversible Plates	Testing Recommended	Best Application	Best Application	Not Recommended
Vibratory Rollers	Not Recommended	Best Application	Testing Recommended	Best Application
Rammax Rollers	Testing Recommended	Best Application	Best Application	Not Recommended

SAFETY AND GENERAL GUIDELINES

As with all construction equipment, there are many safety practices that should be followed while using compaction equipment. While this instructional guide is not designed to cover all aspects of job site safety, we wish to mention some of the more obvious items in regard to compaction equipment. Ideally, equipment operators should familiarize themselves with all of their company's safety regulations, as well as any OSHA, state agency or local agency regulations pertaining to job safety.

Basic personal protection, consisting of durable work gloves, eye protection, ear protection, approved hard hat and work clothes, should be standard issue on any job available for immediate use. In the case of walk-behind compaction equipment, additional toe protection devices should be available, depending on applicable regulations. All personnel operating powered compaction equipment should read all operating and safety instructions for each piece of equipment. Additionally, training should be provided so that the operator is aware of all aspects of operation.

No minors should be allowed to operate construction equipment. No operator should run construction equipment when under the influence of medication, illegal drugs or alcohol. Serious injury or death could occur as a result of improper use or neglect of safety practices and attitudes. This applies to both the new worker as well as the seasoned professional.

Shoring

Trench work brings a new set of safety practices and regulations for the compaction equipment operator. This section does not intend to cover the regulations pertaining to trench safety (OSHA Part 1926, Subpart P). The operator should have knowledge of what is required before compacting in a trench or confined area. Be certain a "competent person" (as defined by OSHA Part 1926.650 revised July 1, 1998) has inspected the trench and follows OSHA guidelines for inspection during the duration of the job.

Besides the obvious danger of a trench cave-in, the worker must also be protected from falling objects. Unshored (or shored) trenches can be compacted with the use of remote control compaction equipment. This allows to operator to stay outside the trench while operating the equipment. Safety first!

SOIL COMPACTION TESTING

Soils had been the most used material for many civil engineering projects (embankments, foundation pads, road bases) especially as the 'fill' material. Whenever soil is placed as an engineering fill, it is nearly always necessary to compact it to a dense state, so as to obtain satisfactory engineering properties which would not be achieved with loosely placed material. Soil compaction is a process of mechanically pressing together soil particles to increase the density by expelling air from the void spaces of the soil.

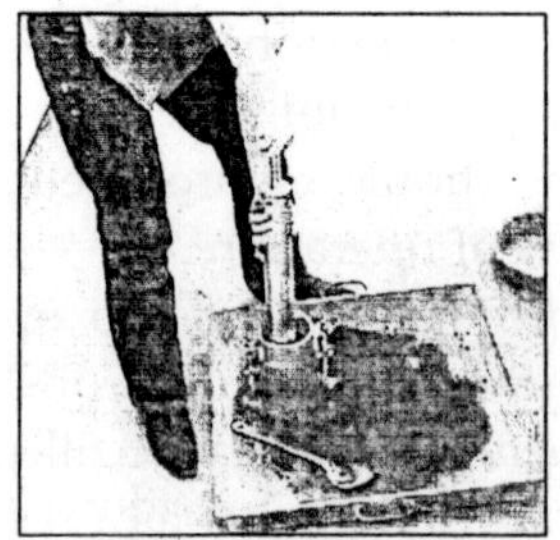

Fig. 5.15 Soil Compaction Test with Rammer and Mould

The compaction of soil is normally performed during construction, often by mechanical means such as rolling, ramming or vibrating.

The degree of compaction control is necessary to achieve a satisfactory result at reasonable cost. Hence, thesoil compaction testing performed at the laboratory shall provide the basis for control procedures used on site.

Soil compaction test shall furnish the following basic data for soils:

- The relationship between dry density and moisture content for a given degree of compactive effort.
- The moisture content for the most effective compaction - that is, at which the maximum dry density is achieved under that compactive effort.
- The achieving value of the maximum dry density (MDD) and optimum moisture content (OMC).

Purposes of Soil Compaction by Machinery:

- Compaction is the application of energy to soil to reduce the void ratio. This is usually required for fill materials, and is sometimes used for natural soils.
- Compaction reduces settlements under working loads.
- Compaction increases the strength of soil material.
- Compaction makes water flow through soil more difficult.
- Compaction can prevent liquefaction during earthquakes.

Objective of Soil Compaction Testing at Lab:

- To determine the dry density -moisture content relationship of a given soil sample using the CBR (California Bearing Ratio) mould and 4.5 kg metal rammer.
- To determine optimum water content and maximum dry density of soil samples.
- To provide the basis for determining the per cent compaction and water content needed in the field.
- To provide guidelines for construction control as to assure that the target values, the degree of compaction are achieved.

Technically speaking, there are several different standard of soil compaction testing for the laboratory. The test selected for use as the basis for comparison will depend upon the nature of works, the type of soil, and the type of compaction equipment used on project site.

- The amount of water content used upon soil samples.
- The type of soil samples being compacted at the lab.
- The amount of compactive energy used in compaction.

Standard Soil Compaction Test:

- Clause D698 of ASTM - Standard test methods for laboratory compaction characteristics of soil using standard effort (12,400 ft-lbf/ft3 or 600 kN-m/m3).
- BS 1377: 1990: Standard soil compaction testing method with 2.5 kg rammer method.

Compaction Test for Soil Brief Explanation

In geotechnical testing laboratory, at least 4 to 5 specimens with water contents bracketing the estimated optimum water content were used in soil compaction testing. A specimen having water content close to optimum would be prepared first by trial additions of water and mixing and then water contents for the rest of the specimens would be selected to provide at least two specimens wet and two specimens dry of optimum, and water contents varying by about 2%, but no more than 4% of water usage.

The obtained data, when plotted, represents a curvilinear relationship known as the compaction curve. The values of optimum moisture content (OMC) and standard maximum dry density (MDD) are determined from the compaction curve diagram. Hence, the soil compaction testing could provide all of that which used for the construction purposes.

HYDROMETER TEST - THE DENSITY OF SOIL

What is a hydrometer analysis for soil?Hydrometer test is the procedure generally adopted for determining the particle-size distribution in the soil for the fraction for that is finer than sieve size 0.075 mm. The lower limit of the particle size determined by this procedure is about 0.001 mm. In soil hydrometer testing, the soil sample is dispersed in water. In the dispersed state in the water, the soil particle will settle individually.

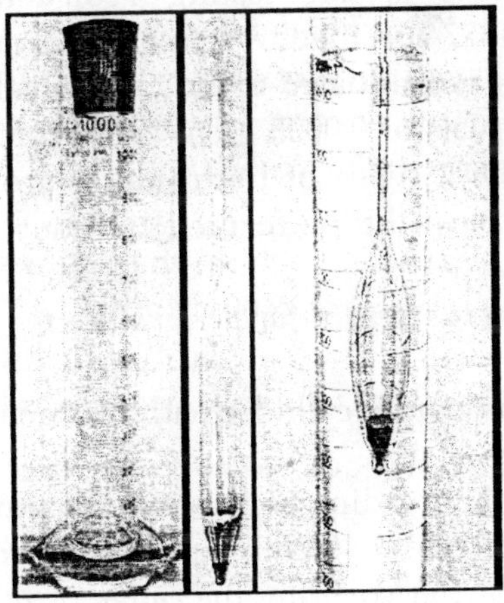

Fig. 5.16 Measuring Cylinder and Hydrometer

It is assumed that the soil particles are spheres, and its velocity can be given by the Stoke's Law. The 152 H type hydrometer of ASTM D422 will be used. If the soil hydrometer is suspended in water in which the soil is dispersed, it will measure the specific gravity of the soil-water suspension at the depth L. The depth L is called as effective depth.

In this method a density hydrometer of special design is used to measure the density of the soil, pre-treated in a suspension in water at various intervals of time. From these measurements the distribution of particle sizes in the silt range (60 to 2 μm) can be assessed. The hydrometer test is not usually performed in less than 10% of the soil material passes the 63 μm sieve pan.

This method can give results which are sufficiently accurate for most engineering purposes. The techniques are less exacting than those required for the pipette method. The hydrometer test method has the additional advantage that it can be performed without much difficulty in a small field laboratory. If the main central laboratory also uses this procedure, the results obtained by both are directly comparable.

Soil hydrometer used for testing of soil density in technical standard of BS1377 specified. An essential requirement is that the scale reading is graduated to indicate density in g/cm³ or g/ml at scale intervals of 0.0005 g/cm³. In technical standard of ASTM D422, the 2 types of hydrometer are specified. The hydrometer test can be performed without any individual calibration.

Other apparatus needed for performing the hydrometer test are:

- Glass measuring cylinder with 1000ml marked and 360 mm high; 2 nos.
- Stop-watch reading to 1 second for time measures.
- Glass rod about 400 mm long and 12 mm diameter.
- Thermometer covering the range of zero to 50°C with reading to 0.5°C.
- Constant-temperature bath capable of being maintained at 25°C ± 0.5°C with deep enough for immersing the sedimentation cylinders to the 1000 ml mark.

The hydrometer test is among the sedimentation theory which based on the fact that large particles in the liquid settle more quickly than small particles, by assuming all the particles have similar densities and shapes.

LIQUID LIMIT TEST - THE LIQUID STATE OF SOIL

The liquid limit test in the laboratory is performed to determine the plastic and liquid limits of a fine grained soil. The liquid limit is arbitrarily defined as the water content, in per cent, at which a part of soil in a standard cup and cut by a groove of standard dimensions.

The soil will flow together at the base of the groove for a distance of 13 mm when subjected to 25 shocks from the cup being dropped 10 mm in the standard Casagrande apparatus operated at a rate of two shocks per second. Liquid limit test is also known as Atterberg limit test.

Fig. 5.17 Cone Penetrometer and Casagrande Apparatus

Whereas, the plastic limit (PL) is the water content, in per cent, at which a soil can no longer be deformed by rolling into 3.2 mm diameter threads without crumbling. The test normally covers the determination of the plastic limit of soil, *i.e.* the lowest moisture content at which the soil is plastic. Additionally, Atterberg defined four possible states of consistency for soils: liquid, plastic, semi-solid and solid.

The liquid limit test shall divide the plastic and liquid states and is defined as the water content at which the soil flows to close a standard size groove when shaken in a standardized device. At this water content the soil has approximate shear strength of 2.5 kPa.

The two main types of test are specified:

1. Casagrande apparatus method, which has been used widely as the basis for soil classification.
2. Cone penetrometer method, which is more satisfactory.

Other Required Equipment for Liquid Limit Test:

- Sieve with 425µm aperture size, Porcelain (evaporating) dish, Flat grooving tool with gage, 8 nos. of moisture cans, Electronic balance, Glass plate, Spatula, Wash bottle filled with distilled water, Drying oven set at 105°C.

The Standards for liquid and plastic limit test:

- D4318 of ASTM - Standard Test Method for Liquid Limit, Plastic Limit, and Plasticity Index of Soils
- Clause 4.3 BS 1377: Part2: 1990 - Cone Penetrometer Test Method for Liquid Limit
- Clause 4.5 BS 1377: Part2: 1990 - Casagrande Test Method for Liquid Limit
- Clause 5.3 BS 1377: Part2: 1990 - Plastic Limit Test

Casagrande Test Method for Liquid Limit

Method is used for determination of liquid limit of a sample of natural soil or of sample of soil from which material retained on a 425 ?m test sieve has been removed. The apparatus are: Liquid limit device and grooving tools, apparatus for moisture content determination, two palette knives, a flat glass plate, a wash bottle, and corrosion-resistant air-tight containers.

Plastic Limit Test

This method covers the determination of the plastic limit of soil, *i.e.* the lowest moisture content at which the soil is plastic. The plastic limit separates plastic and semi-solid states. At water contents below the plastic limit the soil cannot be molded without cracking. The apparatus are: A flat glass plate on which soil is mixed, a flat glass plate on which treads are rolled, two palette knives, apparatus for moisture content determination.

The liquid limit test for ASTM D4318 and BS1377: Part2: 1990 would be in the similar manner and shall use the same required apparatus. It is same for the plastic limit testing of soil samples.

Cone Penetrometer Test Method for Liquid Limit

Method is based on the measurement of penetration into the soil of a standardized cone of specified mass. At the liquid limit the cone penetration is 20 mm. The apparatus are: Penetrometer apparatus, metal cups, moisture content apparatus, metal straight edge, two palette knives. This method also sometimes known as Fall Cone Test.

Nonetheless, the liquid limit, plastic limit, and the plasticity index of soils are used extensively to correlate with engineering behaviour such as compressibility, hydraulic conductivity, shrink-swell, and shear strength.

The 2 most important parameters, LL and PL are obtained from the liquid limit test whereby among the earliest test performed at the Geotechnical laboratory.

CBR TESTING-DETERMINATION OF SOIL STRENGTH

The California Bearing Ratio test or CBR test is meant for determining the soil bearing value or the soil aggregates when the samples are compacted in the laboratory using heavy compaction (4.5 kg rammer) or light compaction (in some cases). The test is useful for evaluating materials containing only a small amount of material retained on the 20 mm sieve size.

The CBR test is performed by punching the standard plunger into the soil specimen at the fixed rate of penetration, and measuring the force required to maintain that rate.

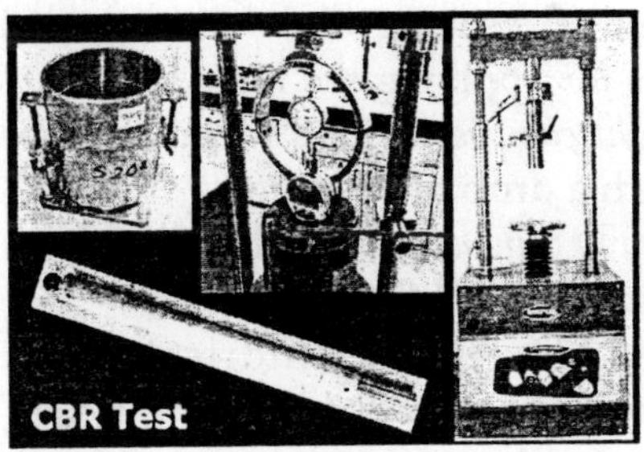

Fig. 5.18 CBR Equipment: CBR Mould, 4.5kg Steel Rammer

The CBR test is the most widely used of the number of empirical penetration type tests. It is perhaps the most adaptable of these tests, for it can be carried out on any types of soil ranging from heavy clay to material of medium gravel size. Moreover, the CBR test is undertaken to determine the strength or supporting value of a given soil if used as a subgrade, subbase or roadbase material. The laboratory test method for the determination of CBR of undisturbed and remoulded or compacted soil specimens, both in soaked (in water for 4 days) as well as unsoaked state.

The technical standards for California bearing ratio test for soil:

- *D1883 of ASTM*: Standard Method of Test for Bearing Ratio of Laboratory Compacted Soils
- *TEST 15 of BS 892*: 1967 or TEST 16 of BS 1377: 1975

The objective of performing the CBR test:

- To determine the California bearing ratio by conducting a load penetration test in the laboratory.
- To determine the relationship between force and penetration when a cylindrical plunger of a standard cross-sectional area is made to penetrate the soil at a given rate. At certain values of penetration the ratio of the applied force to a standard force, expressed as the CBR percentage.

The equipments and tool required for the soil bearing ratio testing:

- CBR machine for applying the test force through the plunger, consisting of a force-measuring device and means for applying the force at a controlled rate.
- *For light compaction test*: Metal rammers of weight 2.5 kg with a drop of 300 mm
- *For heavy compaction test*: Metal rammers of weight 4.5 kg with a drop of 450 mm.
- *For ASTM type*: Cylindrical mould with inside diameter 150 mm and height 175 mm, provided with a detachable extension collar 50 mm height and a detachable perforated base plate 10 mm thick.
- *For BS type*: Cylindrical CBR mould having an internal diameter of 152 mm and an internal effective height of 127 mm with detachable base plate and a collar of 50 mm deep.
- Sieve size of 4.75 mm and 20 mm I.S. test standard.
- Electronic balance capable of weighing up to 25 kg readable and accurate to 5 g.

- Wooden hammer or rubber mallet, Spatula, Apparatus for moisture content determination.
- Miscellaneous apparatus such as drying Oven, the mixing bowl, straight edge, scales soaking tank or pan, filter paper and containers.

The CBR test data are applicable to the design of the roads as well the airfield runways and taxiways. Furthermore, the test also meant for the evaluation of subgrade strength of roads and pavements. The results obtained by these tests are used with the empirical curves to determine the thickness of pavement and its component layers. Hence, the soil CBR test is the most widely used method for the design of flexible pavement.

DIRECT SHEAR TEST - SOIL SHEARING STRENGTH ANALYSIS

Direct shear test is one of the oldest soil strength tests done in the laboratory. The soil shear testing involved the sliding of one portion of the soil specimen on another by using the shearbox test. Hence, the direct shear device will be used to determine the shear strength of a cohesionless soil, such as the angle of internal friction. From the plot of the shear stress versus the horizontal displacement, the maximum shear stress is obtained for a specific vertical confining stress.

Fig. 5.19 Soil Shear Testing Equipment by Means of Automation

The purpose of performing the soil direct shear test:

- This test is performed to determine the consolidated-drained shear strength of a sandy to silty soil.

- The shear strength is one of the most important engineering properties of a soil, because it is required whenever a structure is dependent on the soil's shearing resistance.
- The shear strength is needed for engineering situations such as determining the stability of slopes or cuts, finding the bearing capacity for foundations, and calculating the pressure exerted by a soil on a retaining wall.

The technical standard used for the soil shear testing: Clause D3080 of ASTM - Standard Test Method for Direct Shear Test of Soils Under Consolidated Drained Conditions

The Important Apparatus used for Soil Shear Strength Test:

- *Shear Device*: The device that used for direct shear test holds the specimen securely between 2 porous inserts in such a way that torque is not applied to the specimen. It shall provide a means of applying a normal stress to the faces of the specimen, for measuring change in thickness of the specimen, for permitting drainage of water through the porous inserts at the top and bottom boundaries of the specimen, and for submerging the specimen in water. The device shall be capable of applying a shear force to the specimen in water. The device shall be capable of applying a shear force to the specimen along a predetermined shear plane (single shear) parallel to the faces of the specimen. The frames that hold the specimen shall be sufficiently rigid to prevent their distortion during shearing.
- *Shearbox for Soil Specimen*: Either square or circular, made of stainless steel, bronze, or aluminum, with provisions for drainage through the top and bottom. The box is divided vertically by a horizontal plane into two halves of equal thickness which are fitted together with alignment screws. It is also fitted with gap screws, which control the space (gap) between the top and bottom halves of the shear box.

- *Shearbox Bowl*: The metallic box which supports the shearbox and provides either a reaction against which one half of the shearbox is restrained, or a solid base with provisions for aligning one half of the shear box, which is free to move coincident with applied shear force in a horizontal plane.
- *Device for Shearing the Soil Specimen*: Through shear test, the device shall be capable of shearing the soil samples at a uniform rate of displacement, with less than 65% deviation, and should permit adjustment of the rate of displacement from 0.0001 to 0.04 inch/min (0.0025 to 1.0 mm/min). The rate to be applied depends upon the consolidation characteristics of the soils. The rate is usually maintained with an electric motor and gear box arrangement and the shear force is determined by a load indicating device such as a proving ring or load cell.
- *Device for Applying and Measuring the Normal Force*: The normal force is applied by a lever loading yoke which is activated by dead weights (masses) or by the pneumatic loading device. The device shall be capable of maintaining the normal force to within 61% of the specified force quickly without exceeding it.
- *Device for Shear Force Measurement*: The proving ring or load cell accurate to 0.5 lbf (also 2.5 N) or 1% of the shear force at failure level or whichever is greater.
- *Porous Inserts*: The porous inserts (stones) function to allow drainage from the soil specimen along the top and bottom boundaries. It also functions to transfer horizontal shear stress from the insert to the top and bottom boundaries of the specimen. Porous inserts shall consist of silicon carbide, aluminum oxide, or metal which is not subject to corrosion by soil substances or soil moisture. The proper grade of insert depends on the soil while direct shear testing in progress. The permeability of the insert should be substantially greater than that of the soil, but should

be textured fine enough to prevent excessive intrusion of the soil into the pores of the insert.

- *Miscellaneous Equipment for Direct Shear Test*: Timing device with a second hand, distilled or demineralized water, spatulas, knives, straightedge, wire saws, and etc.

Therefore, the soil shear test is very important in determination of frictional strength of cohesionless soil. The shear testing can be done either in manual mode or in automatic mode. Nonetheless, the soil preparation stage shall be the most concerned.

SOIL CONSOLIDATION TEST - SETTLEMENT OF SOIL DETERMINATION

Consolidation of soil is the process of time-dependent settlement of saturated clayey soil when subjected to increasing loads. Moreover, the consolidation testing is performed to determine the magnitude and rate of volume decrease that a laterally confined soil specimen undergoes when subjected to different vertical pressures. In soil consolidation test, the common test would be the one-dimensional laboratory test.

Fig. 5.20 Automated and Manual Soil Consolidation Apparatus

The measured data obtained from the soil consolidation test, the pressure vs. void ratio relationship of consolidation curve can be plotted. This data is useful in determining the compression index, the re-compression index and the pre-consolidation pressure (or maximum past pressure) of the soil.

In addition, the data obtained can also be used to determine the coefficient of consolidation and the coefficient of secondary compression of the soil.

Dimensional Consolidation Properties of Soils

The Significance of Consolidation: The consolidation properties determined from the soil consolidation test are used to estimate the magnitude and the rate of both primary and secondary consolidation settlement of a structure or an earthfill. Estimates of this type are of key importance in the design of engineered structures and the evaluation of their performance.

The equipment used for ASTM consolidation testing:

- Consolidation device including ring, porous stones, water reservoir, and load plate.
- Dial gauge with reading of 0.0001 inch = 1.0 on dial.
- Soil samples trimming device with glass plate.
- Metal straight edge, clock, moisture can, and filter paper.

Oedometer Soil Consolidation Test

The objective of the soil consolidation test is to determine the magnitude and the rate of soil consolidation when it is restrained laterally drained axially while subjected to incrementally applied loading.

The two parameters to be determined are:

1. The compressibility of the soil (expressed in terms of the coefficient of volume compressibility), which is a measure of the amount by which the soil will compress when loaded and allowed to consolidate.
2. The time related parameter (expressed in terms of the coefficient of consolidation), which indicates the rate of compression, and hence the time-period over which consolidation settlement will take place.

The apparatus commonly used for Oedometer soil consolidation test:

- The soil consolidation cell with porous stones.
- The consolidation loading device and dial gauge.

- Flat glass plate, such as is used for the liquid limit test.
- Cutting tools and straight-edge for specimen trimming.
- Auxiliary items for determining the moisture content.
- Balance and measuring instruments like thermometer, for weighing and measuring the soil specimen and ring.

Understanding the Benefits upon Engineering: When structures are built on saturated soil, the load is presumed to be carried initially by incompressible water within the soil. Because of additional load on the soil, water will tend to be extruded from voids in the soil, causing a reduction in void volume and settlement of a structure. In soils of high permeability (course-grained soils), this process requires a short time interval for completion, with the result that almost all of the settlement has occurred by the time construction is complete.

However, in soils of low permeability (fine-grained soils, particularly clayey soils), the process requires a long time intervals for completion, with the result that strain occurs very slowly. Thus, settlement takes place slowly and continues over a long period of time. The phenomenon of compression due to very slow extrusion of water from the voids in a fine-grained soil as a result of increased loading (such as the weight of a structure on a soil) is known as consolidation. Associated settlement is referred to as consolidation settlement. It is important to be able to predict both the rate and magnitude of the consolidation settlement of structures. By performing the soil consolidation test, the geotechnical aspect of ground can be understand further with settlement could be minimized.

The CBR test data are applicable to the design of the roads as well the airfield runways and taxiways. Furthermore, the test also meant for the evaluation of subgrade strength of roads and pavements.

6

Behaviour Concepts for Petroleum Fluids

INTRODUCTION

Most reservoir engineering, production, formation evaluation, and drilling issues are ultimately reduced to the nature and spatial distribution of rock properties. This observation provides the incentive for this course; about 90% of the insight into such properties can be captured through fairly simple theoretical and mathematical relations.

This course reviews fundamental principles dealing with rock properties and introduces advanced concepts about displacements, residual phase saturations, and effective properties. We will rely strongly on conceptual and simplified mathematical models. As will be emphasized in this Introduction, the entire course will focus on the petrophysical properties that go into the input of a numerical simulator.

OBJECTIVES

When you finish you should:

- Understand permeability and its origins,
- Be able to illustrate capillary pressure and how to use it,
- Find out what factors affect relative permeability,

- Illustrate heterogeneity measures,
- Understand the difference between heterogeneity and correlation, and
- Show the benefits of statistical assignments.

In each case, we will attempt to make some specific points which should be useful in evaluating data, making recommendations about laboratory procedures and in using information to its best extent. As the outline shows, we will concentrate on single-phase and water/oil properties although we will refer to gas/oil and more exotic properties from time to time.

NUMERICAL SIMULATION

Since this course deals with the quantities that go into the input of a simulator, its appropriate to spend a few minutes talking about simulation in general. Most knowledge progresses through the building, testing, and reformulating of models or representations of reality. This statement is true regardless of the field of study.

MODELS – FOUR BASIC TYPES

1. *Conceptual*: e.g., Fluvial deltaic
2. *Physical*: e.g., Hele-Shaw cells, corefloods
3. *Numerical*: State of the art for engineers
4. *Philosophical*: Whether we know it or not

Geologists are quite comfortable with the idea of model building even though their models (like models in many other fields) are rarely numerical. Physical models were very common in the early days of the petroleum industry, but these have been supplanted by numerical models.

Actually, numerical models have had a long pedigree in hydrocarbon production prediction as the following illustrates.

NUMERICAL MODELS – A SHORT HISTORY

- Tank type--Emphasis on fluids
  - Volumetrics
  - Primary depletion

- Streamline Emphasis on patterns and rates, permeability anisotropy
  - Areal sweep
  - Oil in place
- Simulators -- All of the above plus heterogeneity

Tank type models are still useful today in determining primary drive mechanisms, oil and gas originally in place and even in the initial stages of an involved simulation study.

Fig. 6.1

Streamline models,invented in the mid 1950s, have experienced a resurgence in interest because, in part, of the ability to assist in fluid flow visualization and the ability to solve very large problems. The image to the right is a classical illustration of the streamlines in a 5-spot pattern.

This figure 6.1 also illustrates one of the best uses of streamline modeling; identifying the amount of fluid that will pass outside of a given area. What we mean by anumerical model orsimulator in this course is something which represents the behaviour of one system through the use of another.

- First system – Reservoir and process
- Second system – A model

A model or numerical simulator is a sequence of numerical operations whose output represents the behaviour of a particular process in a particular reservoir. It is, therefore, an integrating tool for combining, in their proper weight, all of the factors that influence production in a reservoir. Models are nothing more than solutions to conservation equations that are coupled to several phenomenological laws that are needed to make the number of equations and unknowns equation. L

BASIC EQUATIONS

- Conservation of
  - Mass
  - Energy
- Empirical laws
  - Darcy
  - Capillary pressure
  - Phase behaviour
  - Fick
  - Reaction rates

Even though the number and type of equations can be said briefly, there is enormous complexity in the laws as the image to the right illustrates. The idea here is to impress rather than inform.) In general, there is one partial differential equation to be solved for each component and these must be solved in three spatial dimensions. These equations are far too complicated to be solved analytically (directly). Instead, what is done, when solving for a reservoir, is to divide it up into a number of cells or grid blocks and the equations solved on these much smaller volumes. This division is, of course, arbitrary but it is the only way to proceed. It would be no surprise to find that the division into grid blocks leads to a number of artifacts in the simulator results. But it should equally be of no surprise to realise that these artifacts tend to disappear as the number of blocks increase; with current simulation technology and with newer simulation technology the artifacts seem to comprise less of a problem than simply not knowing the simulator input to begin with.

TYPICAL INPUTS

Fluids

The fluid input, summarized below, ranges from the basic

properties of viscosity, solubility and volume relations, through much more complicated pressure-volume-temperature relationships embodied in and equation of state simulator.

- Viscosities
- Densities
- Surface tensions
- Phase behaviour

Reservoir

The reservoir input deals with properties that define the gross volume of the fluids in a reservoir.

- Depth (top surface)
- Thickness (bottom surface)
- Lateral extent
- Original contacts

Blocks

What this course deals with the most are the block-by-block petrophysical assignments.

- Initial pressure and saturations
- Porosity
- Permeability (three values)
- Relative permeability
- Capillary pressure

Wells

Well properties can be among the most important to a simulation run because it is through these that the operational decisions about how to deplete a field are quantified.

- Location
- Completion interval
- Productivity (e.g. skin factor)
- Status (e.g. producer)
- How operated

The bullet location can include some quite complicated geometries for horizontal or deviated wells. Many times a great deal of the simulator is devoted to modeling effects in the wellbore and it is becoming common for the well model to be hooked up to separate models of surface facilities.

This input, furthermore, constitutes the main way that the geology is input into a simulator. The lack of quantitative detail in the geological description coupled with the sheer volume of data means that this type of input has historically been among the most overlooked in simulator input. This neglect forms one of the prime motivations for this course.

DIFFICULTIES WITH NUMERICAL SIMULATION

Though it is the only one discussed here, the assignment of grid block properties is only one of several difficulties with numerical simulation.

Some of these are:

- Truncation errors
- Stability
- Grid orientation effects
- Property assignments
- Scale adjustment

Property assignments to cells away from measurements (usually at wells) form the culmination of the material in Module 3. The last item, adjustment for scale or scaling up, is truly a simulator data input also. It involves the fact that the size of the blocks in a typical simulation run is invariably much larger than the size of the measurements. The following table illustrates some of the typical grid block sizes.

Table. Typical Grid Block Sizes

Process	DX (Meters)		DZ (Meters)	
	min	max	min	max
Undersaturated oil reservoir	500	500	7	9
Gas reservoir	1500	1500	70	229
Simulation of well test	244	671	6	61
Retrograde gas reservoir	236	457	34	61
Solution gas drive reservoir	610	610	3	18

Where DX and DZ are the lengths of the sides of the grid blocks. From Haldorsen. Rather than speaking of grid blocks we should be speaking of grid pancakes. And the size of these blocks are those of a modest sized building.

NOTATION

This wull adopts certain conventions from that work, particularly regarding subscript conventions of the various fluids.

Table. Subscript Conventions

1	Water or aqueous phase
2	Oil or oleic phase
3	Gas phase
nw	Nonwetting
w	Wetting
r	Residual
R	Remaining

Other symbols are Society of Petroleum Engineers or industry standards. Thus, S_{wr} means the wetting phase residual saturation and S_{2R} means the remaining oil saturation. We will attempt to maintain consistency in definitions and to define each as near to the first usage as possible.

FLUID FORCES

In two-phase fluid displacements there are mainly three type of forces: viscous forces in the invading fluid, viscous forces in the defending fluid and capillary forces due to the interface between them. This leads to two dimensionless numbers that characterize the flow in porous media: the capillary number C_α and the viscosity ratio M.

The capillary number is a quantity describing the competition between capillary and viscous forces. Consider a porous medium of pores with average size a. Microscopically, the pressure drop Δp_u due to viscous forces across an average pore is obtained from Darcy's equation:

$$\Delta p_u = \frac{u\mu a}{k},$$

Where k is the local permeability of the pore, μ is the viscosity of the fluid and u denotes the flow velocity through the pore. Assume that the pore contains an interface between the liquids and that the curvature of the interface is approximately equal to the average pore size. The corresponding capillary pressure Δp_c is proportional to γ/α where γ is the interfacial tension. The capillary number may now be defined as the ratio of the viscous and capillary pressure:

$$C_\alpha = \frac{\Delta p_u}{\Delta p_c} = \frac{u\mu a^2}{\gamma k}.$$

The permeability of the pore is roughly given by α^2 and we obtain:

$$C_\alpha = \frac{\mu u}{\gamma}$$

Thus, the capillary number becomes a ratio of viscous and capillary forces acting at the pore scale. In two-phase flow in porous media the capillary forces are local due to the pore-interfaces while the viscous forces act across the fluids at all length scales. The porous medium we will consider is statistically homogeneous at large length scales but random on the pore scale due to the pore size distribution.

Thus, the local randomness in the structures of the two fluids are caused by the capillary pressure while the macroscopic flow behaviour is dominated by viscous forces. It is not trivial to know how to measure the competition between such forces and recent work has tried to establish a more accurate expression.

For instance, the above definition does not cover important properties such as the pore size distribution and in two-phase flow it is not clear which viscosity should be inserted. Usually, authors tend to use the maximum viscosity of the two liquids under the assumption that the viscous forces are dominated by the forces in the highest viscous fluid.

The viscosity ratio M is defined as:

$$M = \frac{\mu_2}{\mu_1},$$

where μ_1 and μ_2 refer to the viscosity of the defending and invading fluid respectively.

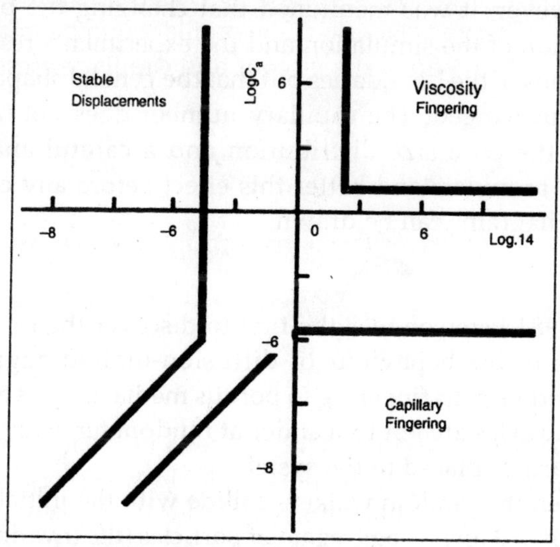

Fig. 6.2 `Phase-diagram" of Lenormand *et al.*

Lenormand's ``phase-diagram" on a logarithmic plot with the viscosity ratio M along the x–axis the the capillary number C_α along the y–axis. The three regions bounded by thick lines corresponds to the three major flow regimes: viscous fingering, stable displacement and capillary fingering obtained in the simulations and experiments performed by Lenormand *et al.*. Note that the capillary number in the ``phase-diagram" is defined by always inserting the viscosity of the invading fluid, even if this is the lower viscous one.

Lenormand *et al.* introduced the concept of ``phase-diagram" in 1988 for drainage displacements where various experiments and simulation were plotted in a plane with the capillary number C_α along the x–axis and the viscosity ratio M along the y–axis. The plot, reproduced in figure 6.2, clearly shows that the different structures they obtained divide into the major flow regimes whose region of validity in C_α and M space is given by the plot. The boundaries of the regions were qualitatively discussed and they concluded that the drainage displacements where fully characterized by C_α and M.

However, it was mentioned that changing the pore size distribution of the simulation and the experiments resulted in translations of the boundaries but that the general shape should remain unchanged. The capillary number does not take into account the pore size distribution and a careful analysis is required to understand better this effect before any complete ``phase-diagram" can be drawn.

DLA

In 1984 Paterson was the first to discover the remarkable parallels in the behaviour of diffusion-limited aggregation (DLA) and viscous fingering in porous media. DLA is a process where particles are left to wander at random far away from an initial particle placed in the system.

When the random walkers collide with the initial particle they stick to it and an aggregate of particles (the invading fluid) is developed. The aggregate grows as new particles stick to the surface of the aggregate to form a structure surprisingly like viscous fingering. In the continuum limit DLA is described by the diffusion equation:

$$\frac{\partial C}{\partial t} = D\nabla^2 C,$$

Where C is the concentration of particles as a function of position and time and D is the diffusion constant. With a steady flux of particle from a source that is far away from the aggregate of sticked particles we obtain $\partial C/\partial t = 0$ resulting the diffusion equation to reduce to the Laplacian:

$$\Delta^2 C = 0.$$

This process is analogous to viscous fingering in porous media when the defending fluid has a much larger viscosity than the invading fluid, *i.e* $M \rightarrow 0$. In this limit we may approximate the pressure gradient in the invading fluid by zero and the fluid flow is described by Darcy's equation applied to the defending fluid only. Thus,

$$U = -\frac{K}{\mu}\nabla p,$$

where U is the flow rate, and Δp denotes the pressure gradient across the defending phase. Combining this with the incompressibility condition $\Delta^2 p=0$ We obtain: $\Delta^2 p = 0$.This is the same equation as above with p replaced by C. In spite of the similarities there is a big difference, though. The disorder of the DLA process is caused by the random walkers (annealed disorder) while the disorder of porous media is inherent given by the pore size distribution (quenched disorder). Thus, there is no simple one to one correspondence between DLA and the viscous fingering process.

Recently, simulations and experiments have shown that the fractal dimension of the structure produced by DLA is similar to the value found in the experiments. In addition, a modified DLA process has been developed to study the dynamics of the front velocity

INVASION PERCOLATION

Over the last two decades there has been much effort to relate the observed structure to percolation theory. Motivated by the study of flow in porous media Wilkinson and Willemsen developed in 1983 invasion percolation, a new form of percolation theory taking into account the fluid transport process. The theory is based on mapping the threshold pressure of each pore to an occupation probability. In an idealized medium the network of pores and throats may be viewed as a regular lattice in which the sites and the bonds represent the pores and the throats. A random number f_p in the unit interval is assigned to each site and bond to correspond the threshold pressure. Thus, f_p is the occupation probability indicating that the actual site or bond is filled with the invading fluid at the capillary pressure corresponding to f_p. The invasion percolation process proceeds by letting the displacing fluid grow each time step by occupying the accessible site along the front having the smallest random number. Regions of defending fluid which become disconnected from the outlet are trapped, and the site of the trapped cluster can not be invaded.

The process is stopped when the invading fluid first percolates, *i.e.* forms a connected path between the inlet and the outlet. The invasion percolation process is only valid in the limit of extremely slow displacement, that means $C_\alpha \rightarrow 0$. In that limit the capillary pressures due to the pore-interfaces are assumed to be in capillary equilibrium and a pore or throat is only invaded by displacing fluid if the capillary pressure is equal to or larger than the corresponding threshold pressure.

Wilkinson and Willemsen measured the fractal dimension of the percolation cluster and they found that the mass M of the invaded fluid increases with the length L of the lattice as: $M(L) \propto L^{Dtrap}$ where, $Dtrap \simeq 1.82$ is the fractal dimension of the cluster taking account of the trapping.

Experiments performed by Lenormand and Zarcone in 1985 confirm this value and the percolation cluster is found to have the same statistical properties as the structures obtained from the experiments. More recently, a modified invasion percolation process has been developed to study the burst dynamics in slow drainage

ACTION

Capillaries play a vital role in the exchange of nutrients and waste products (the primary function of the cardiovascular system). The density of capillaries determines the total surface area available for exchange, the maximum distance between a cell and blood and thus diffusion time, and contributes to the total resistance of the capillary bed.

Alterations in capillary blood flow and pressure may influence the transcapillary exchange processes, the magnitude of this effect being influenced by the lipid-solubility and size of the molecules concerned, and the tissue metabolic requirements. Capillaries also play an important role in tissue fluid homeostasis.

The transport of fluid across the microvascular wall is controlled by Starling's forces the most variable of which is capillary pressure with the effect that small changes in capillary pressure can have marked effects on transcapillary fluid flux].

The regulation of capillary flow, pressure or the movement of molecules across the capillary wall has been discussed in reviews elsewhere Examination of human capillaries is limited by access to this vascular bed and the techniques available. Although one might imagine that blood flow, pressure and the mechanisms controlling blood flow at the capillary level can be deduced by extrapolating from the findings in larger vessels, the information from animal studies suggests that this is not the case.

Thus there is no alternative but to investigate capillary function directly. In man investigations are confined to capillaries of the retina, lip and skin. Most work has been carried out on skin capillaries which will therefore be the main focus of this review. Before describing the methods used to examine human skin capillaries, skin vascular anatomy will be described briefly.

THE ANATOMY OF THE CUTANEOUS MICROVASCULATURE

The skin blood supply originates from perforating vessels rising from the underlying muscles and subcutaneous fat to form a plexus, the lower horizontal plexus, at the dermal-subcutaneous interface. From the lower plexus, paired arterioles and venules rise to form direct connections with a second plexus, the subpapillary plexus which is situated in the papillary dermis, and from this the capillary loops of the dermal papillae arise.

The majority of the microvasculature of the skin resides in the papillary dermis 1-2 mm below the surface of the skin. Microvessels in the papillary dermis range in size from 10 to 35

?m whereas those in the mid to deep dermis are 40-50 ?m with an occasional arteriole as large as 100 mm being observed.

CHANGES IN MICROVASCULAR ANATOMY WITH SITE AND AGE

Microvascular anatomy, particularly the capillary loops, may vary according to the skin area examined and the age of the subject. Uniquely in the toe and finger nail fold, the terminal row of dermal capillary loops lie parallel to the surface of the skin.

Moving proximally along the digit the orientation of capillaries changes to become perpendicular or oblique to the surface. Both orientations of capillaries are also found at other skin sites although the relative numbers vary.

The development of the capillary loop can also vary, for example in forearm skin where the dermal papillae are not well developed the arterioles connect to capillaries which course close to the dermal-epidermal interface before joining a post capillary venule of the subpapillary plexus.

Total capillary density varies according to the area of skin examined and differences over small areas such as the dorsum of the foot have been reported. Ageing is accompanied by a loss in dermal volume, a reduction in capillary density, shortened capillary loops, and rarefaction of larger microvessels

EXAMINATION OF HUMAN CAPILLARIES-A HISTORICAL PERSPECTIVE

Thirty years after William Harvey elegantly first described the blood circulation, the tiny vessels which link the arterial and venous tree were identified. Visualization of blood flow in these vessels, which are similar in diameter to red blood cells, was first made in frog capillaries by Anthony van Leeuwenhoeck and in 1879 the first microscopic examinations of human skin capillaries were conducted.

In 1922 Müller published a book in which his examinations of skin capillaries using microscopy were illustrated in colour by artists, and show the movement of red blood cells in human nailfold capillaries and the morphology of capillaries in several disease states. The initial measurements of capillary blood velocity were made in 1919 by Basher. Subsequent measurements were reported in 1964 by Zimmer and Demis who used a microscope-television system to study the flow dynamics in human skin capillaries.

A new television-microscopy system introduced in 1974 by Bollinger and colleagues showed that it was possible, using a frame-to-frame analysis of the movement of plasma gaps along a capillary, to examine capillary dynamics in healthy controls and patients.

This was improved and simplified by Fagrell *et al* who used a video-photometric cross correlation technique which allowed continuous assessment of capillary velocity over longer periods of time and could, although technically difficult, be used by experienced personnel in the clinic.

Cannulation of human capillaries and direct measurement of capillary pressure were first performed by Carrier and Rehberg in 1923. Even before this time indirect methods to estimate capillary pressure were reported although the values obtained do not agree with direct measurements. In 1930 Landis published his seminal paper on measurement of capillary and venous pressure and the effects of physiological and pharmacological interventions.

His technique involved introducing a micropipette attached to a water manometer into a capillary and adjusting manometric pressure until blood did not enter the micropipette tip.

Manometric pressure at this equilibrium point represented mean capillary pressure. Further early studies using the manometric system in patients with hypertension, heart failure, and glomerulonephritis were published and then interest in capillary pressure dwindled until the late 1970s. The disadvantage of the manometric technique is that only averaged pressure can be determined.

The first dynamic measurements of human capillary pressure were described in 1979 using a servo-nulling system similar to that used today.

CURRENT INVESTIGATION OF HUMAN SKIN CAPILLARIES

Capillary microscopy or capillaroscopy allows visualization of a living system in real time, it may be referred to as intravital capillary microscopy or intravital capillaroscopy although in the clinical setting 'intravital' is often omitted. Capillaroscopy provides a 2-D projection of a 3-D network of capillaries. In combination with television and video and/or computer technology capillaroscopy generates high contrast images of skin capillaries on videotape, computer disc or photograph.

The native technique may be used to assess capillary morphology, capillary density, capillary blood velocity (dynamic capillaroscopy), capillary red cell column width or to facilitate direct capillary cannulation. In combination with intravenous administration of fluorescent dyes, e.g. sodium fluorescein or indocyanide green (fluorescence video microscopy or fluorescence angiography), capillaroscopy can be used to examine heterogeneity of capillary flow distribution, to visualize structures not detectable by native capillaroscopy such as capillary aneurysms, and to follow the transcapillary diffusion of tracer as a marker of capillary permeability to small solutes.

The skin pigment, melanin, absorbs light strongly in the visible spectrum making capillaroscopy difficult in highly pigmented skin. However fluorescence video microscopy can be used to estimate capillary density in these individuals. Furthermore the intravital microscopy technique has been used in combination with subepidermal injections of FITC dextran to examine microlymphatic network function and microlymphatic capillary pressure (using the servonulling system) in human skin.

METHODS

Equipment Requirements for a Capillary Microscope

Clinical examination is facilitated if the microscope can be moved over the skin area of interest rather than moving the patient. To achieve this the optics, on a standard mounting block, can be attached to a focusing block which is itself attached to an arm to allow movement up and down, side to side and preferably tilting in two planes. A very light sensitive black and white camera (*e.g.* Philips CCD camera LDH 0703/30) is mounted directly over the lens system and the skin is illuminated by a 50 or 100 W mercury vapour light (Zeiss) directed to the tissue either through the lens (epi-illumination) or via a fibre optic light guide.

The lens should have long working distances (1-1.5 cm) and preferably an adjustable diaphragm to control light input to the camera. The microscope objective should be at right angles to the skin surface. Illumination is one of the most important parts of the capillary microscope. The emission spectrum of mercury vapour is similar to the absorption spectrum of haemoglobin (370-450 nm), thus red blood cells appear black in the image and it is the movement of these cells which is observed when examining capillaries under the microscope.

A heat filter and other filters, such as blue or green, are also required to enhance the contrast sufficient for analysis. The particular filter requirements will depend upon: the nature of the illumination, the spectral characteristics of the Mercury vapour light (which alter with hours of usage), and the patient's skin characteristics. The images are displayed on a high resolution black and white monitor with final magnification 120-400 fold and stored on sVHS video tapes, computer discs or as photographs. Time and date marks on the images allow subsequent identification of a study.

Verbal annotation of video tapes may be useful. Further details of equipment requirements are published elsewhere see reviews. A complete, freely moving commercial capillary microscope is not available however, CapiFlow and Leica market modified standard microscopes in an attempt to fill the need. Specifications for cameras, and to a lesser extent monitors and lens, change so rapidly that it is always necessary to reassess a variety of combinations whenever a new system is purchased.

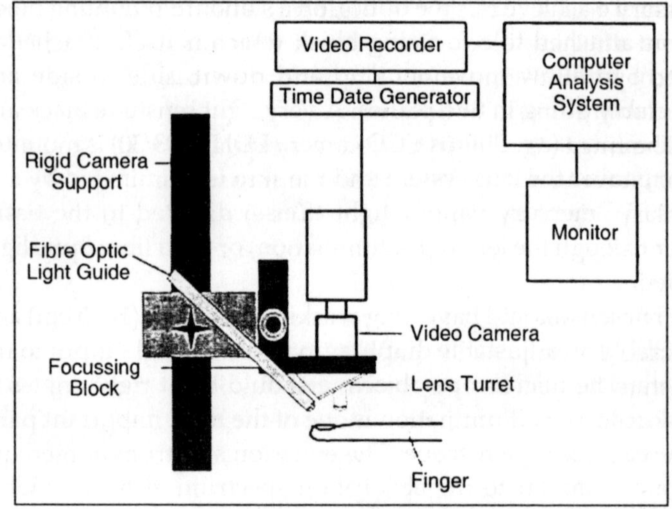

Fig. 6.3

Experimental Conditions for the Examination of Human Capillaries

Due to the thermoregulatory properties of skin and the susceptibility of capillary blood flow and pressure to alterations in venous pressure, sympathetic or sudomotor drive, etc., representative and valid readings for capillary density, red blood cell velocity and pressure will only be obtained if subjects are studied under carefully controlled conditions.

Thus the position in which subjects are examined should be standardized (supine or seated, hand at heart level) and factors such as food intake, smoking, time of day, menstrual cycle, should be carefully controlled. Studies should take place in an undisturbed quiet room with a defined stable temperature, after a period of acclimatization which lasts until skin temperature has stabilized. Skin temperature should be recorded as a variable which influences skin blood flow.

A factor common to all examinations using capillaroscopy is the requirement for excellent images, in particular high black/ white contrast with no movement artifacts. To achieve this it is necessary to immobilize the digit but this must be achieved without affecting blood flow. Blood flow can be artifactually altered by increasing venous pressure, by pressing on the nail, by occluding arterial inflow, etc.

The finger can be supported in a holder moulded to fit. A finger stabiliser attached to the microscope lens may be necessary if long dynamic assessments are being made.

For all applications of capillaroscopy a good image will only be obtained when the considerable scattering of light at the air/ stratum corneum interface is reduced. This can be achieved by applying a thin layer of paraffin oil, glycerine or clear nail varnish to the skin surface. Before the capillaries are examined, time should be allowed for the microcirculation to recover from the perturbation caused by the application of these agents.

The Assessment of Capillary Morphology

Morphological examination of capillaries can be performed by viewing the skin at low power (5-100x) through a simple light microscope or an opthalmoscope. Many but not all morphological assessments have been performed at the finger nailfold. Capillary morphology may be altered by local trauma to the nailfold as may occur in nail biting, or in those engaged in manual work such as builders or farmers.

It is therefore important to note such factors on any assessment forms. When assessments are being made at other skin sites it is important to document the clinical nature of the area since morphology can differ markedly in adjacent areas, *e.g.* avascular atrophie blanche areas compared to adjacent areas of enlarged glomerular like capillaries in chronic venous insufficiency.

The Assessment of Capillary Density

Capillary density is the number of capillaries per unit area of skin. Essentially it is measured by recording images from the capillary microscope and then counting the capillaries in a known area of skin. Depending on the skin area under investigation the capillaries will appear as black dots, if the capillaries are perpendicular to the surface,, or as lines if the capillaries are lying obliquely, or a mixture of the two. It should be noted that the capillary wall cannot be seen and therefore if a capillary is not perfused with red blood cells it will not be visualized under resting conditions using native capillaroscopy.

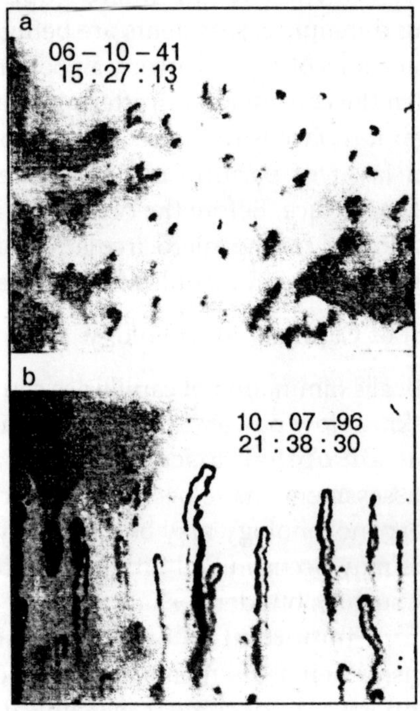

Fig. 6.4

The exact procedure differs from laboratory to laboratory. Four black and white photographs may, for example, be taken around a central point tattooed on the skin and capillaries counted from the photographs. Alternatively six sequential microscopy fields may be recorded for 2 min each (whilst the focusing level is systematically adjusted) and the capillaries counted whilst playing back the video.

Ideally counting should be performed whilst observing the movement of red blood cells and plasma gaps through the capillaries during video replay since this may clarify whether a given black dot is a capillary, whether a knot of vessels is one capillary or two and it will allow the identification of all vessels perfused in the 2 min period rather than those perfused at the instant of a photograph.

It is also possible to combine the two methods described above by counting capillaries from video prints as well as still frame photographs. Recordings of capillary density can estimate the functional density of capillaries under a given condition or the total density of vessels. In the latter case those not perfused at rest must be filled with blood in order to be visualized. This may be achieved either by increasing venous pressure by 40 or 60 mmHg for a few minutes or 2, respectively using a sphygomanometer cuff or by using an arterial occlusion and counting capillaires during the hyperaemia following cuff release.

The disadvantage of the latter is that, in health, the hyperaemia is transient (< 10 s), it is therefore difficult to record the response (subjects move on cuff release, the image needs to be focused at different depths to ensure all the capillaries at different planes have been visualized, etc.). Furthermore the hyperaemia is often followed by a transitory reduction in flow to below resting values which makes it even more difficult to record vessels. The increase in capillary density is around 12-18% in health.

Fluorescence angiography may also be used to identify capillaries which are perfused with plasma only and are therefore not visible under white light. Antonios *et al.* have just performed a direct comparison between arterial and venous occlusion methods of obtaining maximal capillary density and finds venous occlusion to be superior. It is uncertain how much of the increase in capillary density following intervention is due to improved visualization of the capillaires, which appear more black on the image due to the increased number of red cells, and how much is truly due to recruitment of previously unperfused capillaries. It is for this reason that the utmost care must be taken to ensure that all flowing capillaries are recorded during the baseline measurements. Reproducibility of the technique for resting capillary density is 5-8% and for capillary density following intervention is 7%. The use of capillaroscopy in highly pigmented skin is difficult. Native capillaroscopy is only possible in the lighter pigmented areas and these may not be representative of all skin areas. Capillary density can, however, be determined in pigmented skin using fluorescence angiography.

In recent years in addition to determining the numbers of capillaries per unit area of skin further analysis of vessel anatomy have been undertaken. Stereological point counting methods have, for example, been used to assess the vessel volume fraction of skin. This is derived from the Delesse principle that the volume fraction of a component of a three-dimensional solid can be estimated from the area fraction of that component in a two-dimensional state. Such techniques are increasing in popularity but are not as yet widely used in this field.

ASSESSMENTS OF CAPILLARY BLOOD VELOCITY (CBV)

The movement of red blood cells around a capillary is easily visualized under the microscope. The capillary wall cannot be visualized and fluorescent dye studies using indocyanide green have demonstrated that the red cell column width occupies only 67%-68% of the total capillary diameter. True capillary blood flow therefore cannot be quantified by native capillaroscopic techniques.

Capillary red blood cell velocity and an estimate of capillary blood flow (derived using red cell column width as an estimate of capillary diameter) can be obtained but, using the latter to examine the effect of interventions or differences between groups of subjects assumes that the red cell column width remains a similar proportion of the total capillary diameter.

Without care it is easy to subconsciously select the slower capillaries which are most eye catching and are easiest to analyse. To prevent this bias a systematic examination should be undertaken in which recordings start at one side of the nailfold and sequentially record all capillaries until the required number is reached.

Each recording must be long enough to take into account temporal variation in flow, in normal individuals a 2 min recording should be adequate but in some diseased states a longer period may be necessary. A manual estimate of the total time that blood has stopped flowing should be recorded since zero velocity cannot be recorded using the automated systems and to exclude these data would artifically raise the mean CBV for an individual.

As well as measuring resting capillary blood velocity the response (usually time to peak or trough, peak or nadir CBV during intervention, time to recover resting flow) to various interventions can be investigated. This approach may minimize differences between capillaries, standardize the temporal changes in flow between vessels and thus reduce variability between individuals.

Arterial or venous occlusion should be performed by inflating a tiny cuff around the base of the finger rather than the arm as the former maximizes the rate of increase in pressure and improves reproducibility. Responses to local cooling can be investigated as can the effects of the local application of agents by iontophoresis. Each of these procedures requires the assessment of reproducibility by any potential users since considerable variability in responses will occur unless great care is taken.

When used appropriately reproducibility is of the order of 12% for the reduction of CBV seen during local cooling], 18-25% for the hyperaemic peak red blood cell velocity and 19% for the time to peak velocity following 1 min arterial occlusion. Several methods are available to determine CBV. The traditional methods, of frame-to-frame analysis and the flying spot technique have been superceded by automated measurements of CBV using a cross correlation technique.

The principle of this techique is the determination of the time interval by which the upstream window has to be delayed to achieve maximum correlation with the signal from the downstream window, knowing the distance between the two windows, the computer can thereby calculate velocity. Although in principle the technique is easy to use, in practise it requires excellent static images and a considerable amount of user experience to place windows of the correct size, the correct distance apart in order to achieve adequate correlations.

It does not measure zero velocity unless being used in temporal mode when the measurement line is defined along the length of the capillary. The analysis remains very time consuming especially as it is necessary to measure the velocity in several capillaries per subject to obtain a representative value. Its advantages are that it can measure fluctations in red blood cell velocity from moment to moment.

Several other novel methods are also being evaluated for the measurement of CBV such as the spatial shift alignment method. To calculate red blood cell flow, measurement of red blood cell column width is required. In the past this was measured manually using calipers however, computerized analysis using a technique which involves shearing of the vessel image along a line perpendicular to the vessel is now more often used.

Once the vessel has been sheared one of the images is moved until the near edge of its red cell column is aligned with the far edge of the red cell column in the unmoved image. The distance the image has to move to achieve this is equal to the red cell column width.

The Assessment of Capillary Perfusion Using the Capillary Anemometer

This new application of laser Doppler anemometry to measure perfusion in a single capillary offers the opportunity to measure capillary flow at a site other than the nailfold. It involves directing a focused laser beam on to a limb of a single capillary and using Doppler principles to monitor capillary perfusion. It is able to measure greater velocities than can be detected using standard camera and video techniques and clear cardiac pulsatility of capillary flow can be detected especially in high flow conditions. Few data are so far available with the technique although the traditional relationship of capillary perfusion and skin temperature has been confirmed, and marked increases in flow following the application of acetylcholine and abnormalities of capillary perfusion in patients with diabetes have also been described.

The Assessment of Capillary Pressure

Nailfold capillaries can be directly cannulated using glass micropipettes (tip diameters of 5–10 μm) held in a micromanipulator at an angle of approximately 50° to the skin surface. It is usually necessary to remove the dead cornified skin layer to facilitate cannulation. Cannulations at the apex of the capillary loop allow accurate pipette positioning with minimal perturbation to blood flow and thus pressure. Pressure is measured using a resistance null-balance feedback system originally described by Wiederheilm and Intaglietta.

Mean capillary pressure and pre and post cannulation zero values are calculated. An average waveform is obtained by using the R-wave of the ECG to superimpose 10-24 capillary pressure waveforms and the capillary pulse pressure amplitude (CPPA), the time from the R-wave to the foot of the systolic upstroke (systolic arrival time) and the peak pressure are obtained. Fast Fourier transformation may be applied to examine the high frequency components of the waveform.

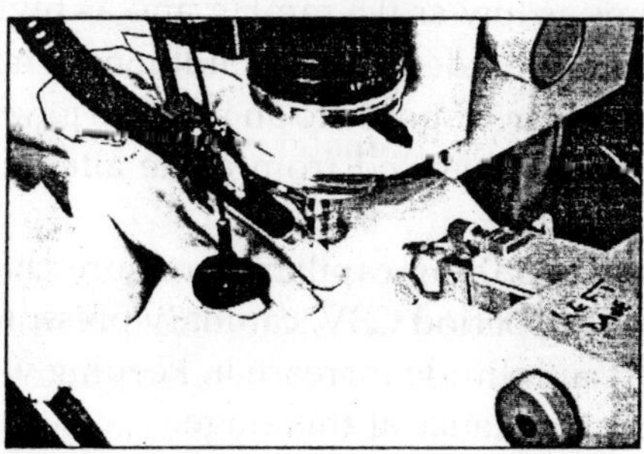

Fig. 6.5

For each subject capillary pressure is defined as the average pressure obtained from a minimum of three capillaries each cannulated at the apex of the capillary loop. In contrast to capillary red blood cell velocity, capillary pressure is a remarkably stable parameter.

Reproducibility of pressure measured in nine capillaries across the nailfold at the same visit is 5.4 ± 2.0% in health and in six subjects studied on three separate occasion the mean coefficient of variation was 5.2 ± 3.6%.

CAPILLAROSCOPY AND CAPILLARY PRESSURE IN HEALTH

The Effects of Age

The reductions in dermal volume which accompany ageing improve visualization of vessels in the subpapillary plexus. There is a striking reduction in capillary density with age and capillary loops are shortened.

Capillary Pressure Physiology

Normal capillary pressure, measured at the apex of the capillary loop with the capillary at heart level, ranges from 10.5 to 22.5 mmHg. It is lower in premenopausal women than in postmenopausal women or in men and does not correlate with brachial artery blood pressure.

Cardiac pulsations are clearly transmitted to the capillary, pulse pressures as low as 0.5 mmHg and as high as 12 mmHg have been recorded under resting conditions in health. Capillary pressure increases in response to an increase in venous pressure, but appears to be protected from acute alterations in arterial pressure.

During local cooling capillary pressure changes little but in the post cooling period CBV, capillary pressure and capillary pulse pressure amplitude increase in keeping with a reduction in precapillary resistance at this stage.

Pharmacological investigations of the healthy capillary bed, particularly those using randomised double blind protocols, are rare. At rest capillary perfusion does not appear to be regulated by nitric oxide, as assessed by the intra-arterial infusion of 1, 2, and 4 μmol min^{-1} L-NMMA for 10 min.

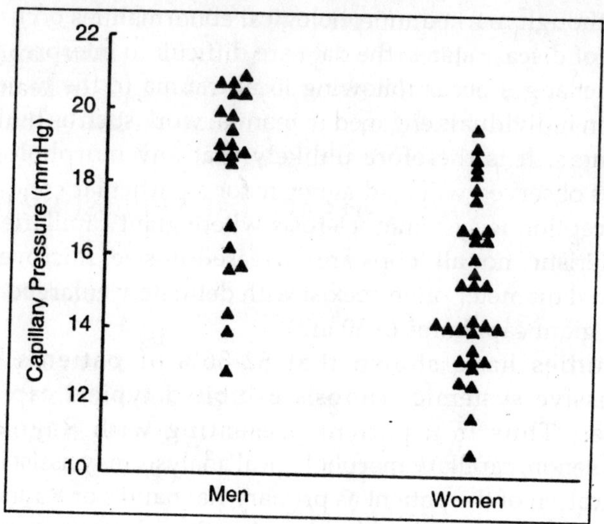

Fig. 6.6 The Effects of Vasoactive Agents on the Healthy Microcirculation

Marked increases in capillary perfusion, capillary pressure and capillary pulse pressure amplitude accompany the nailfold application of the endothelial dependent vasodilator acetylcholine suggesting a reduction in pre to post capillary resistance ratio. However, the mechanisms involved have not been investigated.

Atrial natriuretic peptide (ANP) (5 ng kg^{-1} min^{-1}) augments capillary filtration coefficient suggesting that it enhances extravasation of fluid, and it also has a direct effect on the permeability of the capillary wall to water in vitro.

Recently Houben demonstrated a reduction in nailfold CBV during a 4 h infusion of 7 ng kg^{-1} min^{-1} ANP, which, due to the increased transit time through capillaries and consequently higher plasma osmotic pressure, would tend to offset any water extravasation.

CAPILLAROSCOPY AND CAPILLARY PRESSURE IN DISEASE

Morphological Investigations: A Role in the Clinic

Although marked morphological abnormalities occur in a variety of disease states, the data are difficult to interpret since similar changes occur following local trauma to the hands as occurs in individuals engaged in manual work such as builders or farmers. It is therefore unlikely that any morphological changes observed will be diagnostic for a particular condition. The exception is systemic sclerosis where giant capillaries are characteristic, not all loops are increased in size, microvessels of normal diameter often coexist with definately enlarged (> 20 µm) or giant capillaires (> 50 µm).

Studies have shown that 82-86% of patients with progressive systemic sclerosis exhibited typical capillary changes, Thus in a patient presenting with Raynaud's phenomenon, capillary morphological analysis may assist in the classification of the patient as primary Raynaud's or Raynaud's secondary to systemic sclerosis. Another disease in which capillary morphological analysis has proved useful clinically is in peripheral arterial occlusive disease.

Capillary Density, Red Cell velocity and Pressure in Disease states

This is by no means an exhaustive list of all the diseases investigated.

Diabetes

Skin capillary density is not altered in type 2 diabetes, or subjects with impaired glucose tolerance compared with age, sex and BMI matched controls, and in the kidney no decrease in glomerular number was found in either type 2 patients or type 1 patients with normal renal function.

Capillary Pressure

Capillary pressure is increased in patients with type 1

diabetes, particularly in those with poor glycaemic control or those with incipient nephropathy suggesting that it may be a factor in the pathogenesis of diabetic microangiopathy. In normotensive normoalbuminuric type 2 patients capillary pressure is not raised. However, there is a good correlation between capillary pressure and systemic blood pressure suggesting abnormalities in the normal regulatory processes.

In Type 1 diabetes, abnormalities of skin capillary flow autoregulation have been described within the first year of diabetic life. Such abnormalities are more marked with increasing disease duration particularly when accompanied by poor glycaemic control. Even in those patients with long disease duration without any evidence of clinical complications some degree of microvascular abnormality appears inevitable.

Abnormalities of capillary perfusion may be more marked in the feet than in the hands. Jorneskog and colleagues suggest that abnormal capillary regulation (*e.g.* toe nailfold CBV is severely reduced during reactive hyperaemia in the diabetic group) may contribute to the higher risk for the development of ischaemic foot ulcers in diabetic patients with peripheral vascular disease compared with nondiabetic peers. One hypothesis for the formation of neurogenic ulcers suggests a 'capillary steal' might be involved however, this now seems unlikely as capillary flow is increased not reduced in the diabetic neuropathic foot.

In type 2 diabetes the capillary post occlusive reactive hyperaemia is impaired within the first 10 years after the presentation of diabetes and impairment is more marked in those with poor control.

Treatment Effects

Replacement of human C-peptide in Type 1 patients leads to a redistribution in skin microvascular blood flow by increasing nutritive CBV relative to subpapillary arteriovenous shunt flow . Insulin infusions, in the absence of changes in blood sugar, increase total capillary flow in patients with type 1 diabetes.

Low molecular weight heparin improved healing of diabetic foot ulcers and in seven of the eight patients improved nutritive capillary flow to the ulcer margin. In a randomised double-blind study in healthy volunteers and type 1 patients enalapril failed to reduce capillary pressure whereas in nephrotic type 1 patients an antihypertensive dose of captopril was associated with a reduction in capillary pressure.

Hypertension

Capillary Density

Capillary rarefaction has been described in nailfold and forearm skin, and appears to be a structural rather than simply a functional defect. Similar findings have been described in the conjunctival microcirculation and in other tissues. This structural rarefaction precedes the development of hypertension being present in the dermal vessels of young men who have a familial predisposition to high blood pressure.

Capillary Pressure

Using dynamic measurements of capillary pressure in untreated patients compared to matched controls, Williams reported an elevation in capillary pressure in the hypertensive group thus although increases in peripheral resistance may serve to protect the capillary bed from perturbations of pressure, the increased postcapillary resistance and compliance probably contribute to the elevation in the capillary pressure. Capillary pressure has not been measured in subjects at risk of development of hypertension.

Capillary Blood Velocity

Capillary blood velocity may be reduced or normal in patients with untreated hypertension. Some workers find an inverse relationship between CBV and ambulatory blood pressure, higher blood pressure being associated with a low CBV. The neurogenically induced reduction in CBV may be normal or enhanced.

Treatment

Few studies have examined the effects of antihypertensive medication on capillary function as assessed by capillaroscopy. Neither the ACE inhibitors cilazapril or enalapril, the calcium antagonist amlodipine nor the selective T-type calcium channel blocker mibefradil changed resting CBV significantly in patients with hypertension. This is in contrast to the 40% increase in CBV seen 10 min after the sublingual administration of nifedipine to healthy controls.

Twelve weeks treatment with enalapril did lower the immediate fall in CBV with local cooling but this effect was not sustained over the entire post cooling period, and the authors suggest this finding is probably not clinically significant. In a comparison of carvedilol (α_1/β-adrenoceptor blocker) and bisoprolol (β_1-adrenoceptor blocker), carvedilol increased capillary flow and skin oxygen tension in healthy smokers and in hypertensive patients although this effect was not long lasting.

Peripheral Arterial Occlusive Disease (PAOD)

Capillary Density

In patients with mild disease, the capillary bed of the forefoot does not seem to be markedly abnormal however, with the advent of moderate or severe occlusion profound abnormalities of the skin capillaries have been reported. Fagrell and Lundberg reported a classification system for foot capillaries which was predictive for the development of necrosis and was much more sensitive than systolic blood pressure at the toe.

The final stage (Stage C-a marked reduction in capillary density or no visible capillaries), has a very high predictive value (96%) for the risk of developing skin necrosis in an area over a 3 month period. It has enabled the mapping of a foot to indicate areas 'at risk'.

Lamah and colleagues recently confirmed a reduced capillary density on both feet in subjects with unilateral rest pain or ischaemic ulcers compared to those with intermittent claudication.

Capillary Pressure and Capillary Blood Velocity

Capillary pressure has not been assessed in the foot of patients with PAOD. The post occlusive reactive hyperaemia for CBV is markedly impaired in the PAOD group, however CBV at rest is similar to controls. As total skin flow is increased a maldistribution of flow in favour of non-nutritional skin vessels as opposed to nutritive vessels has been suggested. In healthy capillaries rhythmic variations of CBV are observed, so called flow motion.

In both the fingers and toes of patients with PAOD flow motion is decreased although when present the frequency is similar. Capillaroscopy with fluorescent dyes has been utilized a great deal in PAOD. Such studies have revealed increased transcapillary diffusion of NaF and inhomogeneous microvascular perfusion. In critical ischaemia, CBV and total skin perfusion are reduced and there is a total lack of reactive hyperaemia.

In combination nailfold capillaroscopy, transcutaneous oxygen pressure and laser Doppler perfusion measurements measured at rest and following reactive hyperaemia appear useful in detecting nonreconstructable critical ischaemia that requires amputation which is not detected by blood pressure or clinical indicators. In their study 83% of patients with capillary density < 20/mm^2, absent reactive hyperaemia and transcutaneous oxygen pressure less than 10 mmHg required amputation in the next 12 months.

Treatment

Pharmacotherapy may increase the number of blood filled capillaries per unit area of skin. Fagrell showed this to be the case using PGE1 and there was good agreement between the disappearance of rest pain and the increased capillary density. Spinal cord stimulation can also improve capillary density and microcirculatory reactive hyperaemia.

In a further study the skin microcirculation was examined before and after treatment with either arteriography or percutaneous transluminal angioplasty (PTA) in matched groups of patients. Nutritional (capillary) flow was increased in the arteriography group but the increased blood flow in the PTA group was primarily non-nutritive.

DISCUSSION

Capillaroscopy has been used successfully to monitor skin microcirculatory changes in health and disease. Although until recently it has primarily been a research tool, its use in the differential diagnosis of some diseases (*e.g.* systemic sclerosis) has long been acknowledged.

It's use in the assessment of treatment for peripheral arterial occlusive disease demonstrates it capabilities and simple challenge response testing (*e.g.* reactive hyperaemia) is becoming increasingly common as a clinical tool.

To date capillaroscopy and the measurement of capillary pressure have not been used a great deal in pharmacology. The improvements in automated analysis of capillary red blood cell velocity and capillary density, the novel drug delivery systems which allow drug delivery without perturbation of the local skin microcirculation, the increasing availablility of specific blockers of the pathways involved in the control of the microcirculation, and the growing recognition of the importance of the microcirculatory bed for tissue health make it likely that this will be a rapidly growing area in the future.

The measurement of capillary pressure is likely to remain a research tool. However in combination with simultaneously measured capillary flow, the local application of pharmacological agonists and antagonists, and the introduction of ion selective, O_2 or NO microelectrodes, unique data regarding the control of capillary pressure in man may be obtained. Normal capillary function is essential for tissue homeostasis. Other microcirculatory techniques measure skin perfusion but none is able to monitor the nutritive component passing through capillaries.

Direct monitoring of a capillary using capillaroscopy provides a unique noninvasive method of investigating capillary function in health and in disease states. Furthermore the recent observation that skin melanoma microvasculature is visible using capillaroscopy, and may provide an in vivo tumour microcirculation model, raises the exciting prospect of looking at tumour vascular control and the effects of anticancer therapy on angiogeneis and vascular reactivity, in vivo.

7

Reservoir Rocks

INTRODUCTION

Reservoirs are the porous and permeable rocks that contain commercial deposits of hydrocarbons. Porosity and permeability are the most important physical properties of these reservoirs, so we'll quickly discuss ways of describing these attributes, and how they may be modified by diagenetic changes. Next we'll talk about their size and lateral and vertical continuity, which will be important for calculations of oil and gas reserves

POROSITY

Definition

Porosity is the opening in a rock.

Strictly speaking it is a ratio:

- Porosity = Open space/total volume of the rock plus opening
- Often multiplied by 100 to be expressed as a percentage.
- symbol is the greek letter phi ϕ

Morphological Classification

There are three types of morphologies to the pore spaces:

1. *Caternary*: where the pore opens to more than one throat passage

2. *Cul-de-sac*: where the pore opens to only one throat passage
3. *Closed pore*: where there is no connection with other pores

Morphological types (1) And (2) Are called effective porosity because they allow hydrocarbons to move into and out of them. Morphological type (3) Isineffective porosity becaues no HC can move in or out.

Caternary pore are the best type for a HC reservoir because the HC in the reservoir can be flushed from rocks (*e.g.* secondary and tertiary recovery through water/gas flooding). Cul-de sac pores cannot be flushed, but they can produce HC by release of pressure, *e.g.* gas expansion. Closed pores cannot produce any type of HC.

Thus it is very important (and also very difficult) to determine the ratio of effective to total porosity. Total porosity is fairly easy to measure, but may not tell much about the amount of HC that can be produced.

POROSITY MEASUREMENTS

Can be measured using three techniques:

1. *Well logs*: One of the primary uses of well logs. Often map out subsurface trends of porous lithologies as a play strategy. Different types of logs can be used for porosity measurements depending on lithology.
2. *Seismics*: Through change in acoustic impedence (v*). Density decreases with increasing porosity
3. *Direct measurments of cores*: Generally involves filling pore space with gas, measuring the volume of the gas, and independently measuring the volume of the rock and porosity.

Genetic Classification of Porosity

Standard sedimentologic description of porosity: Primary-formed when sediment deposited.

Two types:

1. *Interparticle*: Lost quickly in muds and carbonate sands

through compaction and cementation respectively. Retained and common in siliciclastic sands.

2. *Intra particle*: Interiors of carbonate skeletal grains.

Secondary- formed after deposition:

- *Dissolution*: Typical of carbonate reservoirs
- *Fracture porosity*: Typically not voluminous, but are very important becuase they can enhance permeability (*e.g.* open closed pores).
  - Fractures can be identified on well logs, seismics, and production statistics, although all identifications are open to interpretation.
  - Fractures are often tectonic related: (1) folds, (2) faults. Also may occur through hydrofracturing where pore fluid pressures are greater than lithostatic.
  - Fractures can create reservoirs out of otherwise impermeable rocks such as basement rocks, or extremely tight sedimentary formations (*e.g.* Bredehoeft article, also the Monterey Fm).

PERMEABILITY

Hydraulic Conductivity

Permeability is a measure of the ability of fluids to pass through a porous medium. This is a more difficult variable to measure in reservoir rocks, but in many ways may be more important than porosity.

Definition, From Darcy's experiment. Darcy flowed water through sand filled tube and found that the flow rate. Fig. 6.6 Fetter Darcy showed that discharge, Q, is proportional to the difference in the height of water, H, and inversely proportional to the length of the tube, L:

$$Q \propto H1\text{-}H2 \text{ and } Q \propto 1/L$$

Discharge is also proportional to the cross sectional area of the pipe, A. In order to make the proportionalities equal, there has to be a proportionality constant, called Kc:

$$Q = -KcA(H1-H2)/L$$

or for infintesimally small segments of the pipe (*e.g.* calculus):

$$Q= -KcA(dh/dl)$$

The deriviative, dh/dl, is called the hydraulic gradient, and the negative sign indicates that the direction of flow is in the direction of decreasing hydraulic gradient. The proportionality constant, Kc, is called the hydraulic conductivity or coefficient of permeability. It has units of velocity, or L/T.

Intrinsic Permeability

Fluid Variables

The hydraulic conductivity is contolled only by the reservoir parameters, but there clearly are controls by the properties of the fluids, specifically its viscosity, μ and its specific weight,γ

Viscosity is a measure of the resistance of a fluid to flow- *i.e.* how much shear force is required to cause the fluid to start to move and continue moving. There is an inverse relationship between viscosity and discharge:

$$Q\alpha 1/\mu$$

Specific weight is the force of gravity on a unit volume of fluid- *i.e.* the force that drives fluids. Discharge is directly proportional to specific weight:

$$Q\alpha\gamma$$

Other Reservoir Variables

Empirical evidence shows that discharge is also proportional to the square of the diameter of the grains that make up the reservoir:

$$Q\alpha d2$$

All of these proportionalities allows a different version of Darcy's law to be written, with a new proportionality constant, C:

$$Q = -(Cd2\gamma/\mu)(dh/dl)$$

Now can introduce a new variable, called intrinsic permeability, K, which represents a propoerty of the porous reservoir only, specifically the size of the pore openings. The relationship between the intrinsic permeability and the hydraulic conductivity is:

$$Kc = K(\gamma/\mu)$$

Intrinsic permeability has units of length squared (L2) or area. It is essentially a measure of the surface area of the grains in a porous medium. The correct (ie SI) units are meter squared (m^2). It common unit in petroleum engineering is the darcy which is defined as: $(1cP)(1cm^3/sec)/(1cm^2)/(1atm/1cm) = 9.87 * 10^{-9}\ cm^2$ Most petroleum reservoirs have permeabilities less than 1 Darcy, so they are reported as millidarcy (md).

Measuring Permeability

Although permeability is one of themore important aspects of a reservoir, it is difficult to measure. Can be measured in drill holes with Drill Stem Tests (DST) or estimated with log response. Also can be measured directly on core samples with Permeameters.

Wire Line Logs

More details later, but in general, the technique relies on changes in log response as a result of infiltration of the drilling fluid. These are not quantitative measurements of permeability.

Permeameters- direct measurements of permeability:

- Devices that hold samples of material to be measured. They can be rocks, or unconsolidated sediment.
- They all have the same mode of operation- they force fluid through the sample and the discharge is measured, form which permeability is calculated.

Interpretation of Permeability

Darcy's law is commonly used to describe flow in the

subsurface. But there are many qualifications to its use. It is formulated for very specific cases.

Also direct measurements of permeability can be difficult:

- There can be no chemical reactions between rocks and fluids (*e.g.* no hydrocarbon generation, cementation, dissolution)
- Darcy's law only applies to porous flow, with uniform porosity. Dual porosity systems, such as vugs, fractures etc are not modelled by darcy's law.
- Direct measurement of permeability can be difficult because of contamination of cores from the drilling mud.
- Permeability is rarely uniform within a reservoir, and often is different in vertical and horizontal directions. Often is controlled by bedding.
- Only one phase fluid can fill the reservoir to use Darcy's law. For hydrocarbon provinces, this is almost never the case- there is commonly water, gas, and liquid HC.

This last problem relates to the distinctly non-linear behaviour between different fluids in the reservoir and leads to ideas of relative permeability, wettability, and capillary pressure. These concepts are very important because they ultimately control the amount of HC that can be produced from the reservoir.

Relative permeability

The ratio of the effective permeability to the total permeability of the rock. The effective permeability is the permeability of one of the phases at some saturations.

Wettability

Wettability is describes the relative adhesion of two fluids to a solid surface. Wettability is controlled by the particularly minerals exposed to the fluids, chemical constituents in the fluids and the saturation history of the samples.

There are different degrees of wettability:

- *Water wet reservoir*: Water will fill the small pores and spread over the surface of the grains. When additional water is added to the system, the reservoir will take up the water, displacing HC.

- *Oil wet reservoir*: Oil fills the smaller pores and coats the grains. Thus the reservoir will take up oil, but not additional water.
- *Other wettability terms*:
  - *Neutral wetatability*: When a reservoir has equal amounts of water and oil coating the surface of the grains
  - *Fractional wettability*: Whendifferent portions of the reservoir water and oil wet to different extents.
  - *Mixed wettability*: Smaller pores are occupied by water, and the larger pores are occupied by oil.

Wettability is very important for all types of fluid-solid interactions: capillary pressure, relative permeability, electrical properties, irreducible water saturation, residual oil and water saturation. Wettability can be difficult to analyse in cores because its characteristics can change during drilling and core-handling.

Capillary Pressure

Capillary pressure results from molecular attraction of liquids and solid surfaces. The phenomenon results in movement of liquids up capillary tubes.

It also controls the convexity of two fluids:

1. Oil wet systems, the oil surface is concave relative to the water, *i.e.* the oil is attracted to the solid surfaces
2. Water wet systems, the water surface is concave relative to the oil.

Capillary pressure can be measured on reservoir rocks.

This is an important parameter of these rocks for two reasons:

1. The reservoir was originally saturated with water, and the oil had to get into the pores somehow, and
2. The oil has to be gotten out, and its capillarity controls the pressure required to get it out and ultimately, how much can be removed.

Test involves taking a plug of the sample and injecting it with some fluid (water, mercury, oil). The presure at which the fluid invades the reservoir is called thedisplacement pressure. The saturation at which no more water can be put into the reservoir is called the irreducible water saturation. There can be different qualities of reservoirs, depending on the pressures required to move the fluids.

Relationship between Porosity, Permeability, and Texture

Texture is a description of the shape, size, sorting and fabric of sedimentary rocks. All of these variable influence porosity and permeability, although there are no clear correlations between the variables.

There is little theory, and empirical evidence in equivocal:

- *Grain shape*: Porosity (and possibly permeability) may decrease with sphericity and rounded grains.
- *Grain size*: Porosity is theoretically independent of grain size, but there is a general empirical correlation between porosity and permeability. May be caused by increased cementation or because of poorer sorting. Permeability decreases with decreasing grain size because pore throats are smaller and the capillary pressure goes up.
- *Packing*: Porosity (and permeability) will decrease with tighter packing. Most reservoirs are buried and altered, so packing is generally not an issue- the rocks are already packed.
- *Deposition process*: No clear relationship, too many other variables
- *Grain orientation*: Controlled primarily by layering in the beds.

DIAGENESIS

The changes to reservoirs that occur following deposition are probably more important than any primary depositional texture, except perhaps sorting. Clearly diagenesis in sandstones differs from diagenesis is carbonates.

Sandstones

- Porosity in sandstones decreases with depth, like all other sediments. The rate at which it decreases depends on the type of sandstone
  - Mature sands retaining porosity longer than immature sands.
  - Poorly sorted sands (with clay minerals) will compact faster than well sorted sands.
- Other factors effecting porosity gradients are the in situ P and T:
  - Higher geothermal gradient cause faster diagenetic reactions, thus faster cementation
  - Abnormal pressure gradients may preserve porosity by reducing compaction.
- Hydrocarbons can preserve porosity- they prevent circulation of water, thereby stopping cementation.

Carbonates

Much of the generalities of sandstone reservoirs can also be applied to carbonate reservoirs. For carbonate reservoirs, their depositional environment can play a very big role in their quality

Reefs

Reefs typically have very high porosities after deposition because they are generally cemented in place. Reefs are also subject to shallow burial diagenesis because they commonly are near sealevel, and the movement up and down of sealevel can cause cementation. Deeper burial of cementation also will occlude porosity. It is important for preservation of reefs that HC migrate into the pore spaces early in order to preserve the porosity.

Carbonate Sands

Carbonate sands have high initial porosity, but the lose the porosity quickly by compaction and cementation with burial. The earlier the cementation, the better the reservoir, because the sands don't lose porosity by compaction.

Carbonate Muds

These deposits are commonly aragonite, which is metastable at earth surface T and P. Its recrystallization to calcite commonly destroys porosity. Some deposits of carboante muds are calcite- these are called chalk, and are typically composed of the tests of coccolithifers.

They can have very high porosities that are preserved with depth of burial. They also tend to have very low permeabilities because of their small size. When they are fractured, they can make good reservoirs. *e.g.* the Austin Chalk of Texas. Very low production rates, but very long lived wells.

Dolomite

Dolomites often form good reservoirs. The common dogma is that it is because Mg is 13% smaller than Ca, so that during dolomitization, there is a total decrease in volume of the material by 13%, thereby generating 13% porosity. A similar argument is made for the conversion of aragonite to calcite, ony now it is the opposite direction and there is not an exchange of elements, only of structure. Enhancement or destruction of porosity during conversion from one mineral to another is unlikely to relate to the mineralogical differences between the minerals. It is more likely to be dissolution or precipitation phenomenon.

Other Types of Reservoirs

These type of reservoirs account for around 10% of the world's production. The remaining 90% is in sandstones and carbonates. Two major processes can create porosity and permeability of other types of rocks:

Dissolution

This is a fairly typical way to generate reservoirs. Largely from dissoltuion of feldspars in granites- called granite wash. This is what the Hugoton field produces from.

Fracturing

Can generate porosity and permeability in otherwise tight

rocks. Fractures commonly vertical, so deveated wells can be useful for drilling these types of reservoirs.

RESERVOIR CONTINUITY

The most important aspect of a reservoir is its dimensions, vertical and horizontal. Much of development geologists jobs involve mapping in the subsurface the dimensions of reservoirs.

Horizontal dimensions are controlled by the depositional environment of the reservoir: *e.g.* barrier island, point bars, reef, sand dunes, diagnesis Vertical dimensions are divided into gross pay and net pay.

- *Gross pay*: The total thickness from the top of the reservoir to the oil/water interface.
- *Net pay*: The total thickness from which the HC can be produced.

The difference between the two is controlled by the porosity distribution within the reservoir. In other words, reservoirs are not continuous- there continuity can be disrupted by barriers to porosity/permeability caused by variations in depositions, subsequent diagenesis or by structures

Depositional Barriers (Sandstones)

Depositional barriers are a result of the shape that reservoir bodies are deposited. These are usually discussed in terms of sandstones, and the sandstones generally have horizontal distributions that are the same as when they were deposited.

There are psuedo-statistical descriptions of the sands:

- *Horizontal dimensions*: Based on their length to width ratios. Given names based on the valuc of the ratios: Sheet, Pod, Ribbon etc.
- *Lateral dimesions*: Often describe on the basis of the thickness and continuity of sandstones. Also can be described on the basis of the continuity of the shales. This is very important for determining the best way to produce the field- *i.e.* exactly where to drill production and flood wells.

Diagenetic Barriers (Carbonates)

- These are largely cemented regions within carbonate reservoirs. They may be zones of mixed waters that allow carbonate precipitation, or oil water contacts.

Structural Barriers (Faults)

- Faults can be open and permeable for fluids- in this case they may connect reservoir and source beds.
- Faults can also be impermeable and thus act as a seal and/or trap for a reservoir.
- Wether the fault is permeable or impermeable depends on may factors: physical properties of rocks, fluid pressure, juxtaposition of sands and shales.

Reservoir Characterization

- Once the barriers have been defined, the next tack is to construct maps of the reservoir that show as much detail as possible.
- Details include, porosity distribution (barriers) and permeability distributions. Information for this comes from a variety of sources: well logs, cores, seismics, production tests, DST's, outcrop information etc.
- It is all of this information that goes into the reserve calculations:

RESERVE CALCULATIONS

These calculations are usually done by a reservoir engineer (if there is one in the company you work for). Small companies may not have engineers, and it is up to the geologist to make these calculations. They are simple and fun.

Preliminary Volumetric Reserve Calculations

- These are the kind of calculations you would go through if you wished to "join" in the drilling or a well, or if you were planning to bid on acreage. Essentially a way to evaluate the value of particular area:

- The technique is to map out the area of the reservoir, contour its thickness and sum up the thicknesses. Then two fudge factors have to be applied:
- Estimated porosity of the reservoir
- *Recovery factor*: depends on drive mechanism, viscosity of oil, permeability, well spacing. Recovery factors can range up to 30% or more for sandstones, most carbonate reservoirs, it will be in the 10 to 20% range.

Postdiscovery Reserve Calculations

Once accurate resevoir data are know, it is possible to make better calculations. This is often done during "unitization" of a field, a process where all owners pool their interest and divide up the recoverable oil. The recoverable oil is calculated according to a formula:

Recoverable oil (bbls) = 7758*V*ϕ*(1-Sw)*R*FVF-1

- The value 7758 converts from acre-feet to bbls
- V = volume of reservoir in acre-feet
- f = porosity
- Sw = water saturation
- R = recovery factor- this can only be estimated, but may be accurate if enough history of surrounding wells is known
- FVF = formation volume factor

The FVF is a scaling factor that converts the volume of oil at T and P or the reservoir to the T and P of the surface, as well as extracts the volume taken up by reservoir gas. it is genreally a number between 1.1 and 2, but since it is in the denominator it reduces the ultimate recovery.

- It can be calculated on the basis of the Gas-Oil ratio
- It can be determined in the laboratory

Note that as the field is produced, the composition of the fluids can change considerably. The primary changes are that the oil/water contact will move up, and a gas cap may form on the top of the oil layer.

PRODUCTION METHODS

The type of driving mechanism that allows HC to flow to the surface is usually the concern of the petroleum engineer. But it is important for a petroleum geologist to be aware of the different kinds of driving mechanism. Three main types: water drive, gas drive, and dissolved gas drive:

Water Drive

In this case, the pressure in the reservoir is hydrostatic. The pressure depends on the recharge at the earth's surface. In this case, oil water contact may rise as the reservoir is produced. The oil water contact in most cases will not rise evenly-controlled by permeability- this will cause fingering of water into the more permeable layers. Also, the water may cone up into the wellbore if the well is produced too fast

Production history: As these reservoirs are produced, reservoir pressure drops inversely with the recharge of the aquifer. There is little change in gas/oil ratio (GOR). The amount of fluid produced generally remains constant, but the water/oil ratio increases.

Gas Cap Drive

In this case, the reservoir contains a gas cap- freee gas sitting on top of oil. During production, the pressure in the reservoir decreases, and causes the gas to come out of solution of the oil. Thus, free gas fills the voids that are left by the produced oil. Production history: Pressure and oil production decrease steadily during production, but the GOR increases.

Dissolved Gas Drive

In this case, there is sufficient gas in solution to keep pressure of the reservoir high. The expantion of as it comes out of solution keeps the pressure high. If the pressure drops low enough, free gas cap will form. This is called the critical gas saturation. At this point, pressure decreases quite a lot, and production drops off.

Artificial Lift and Enhanced Recovery

- If original pressure is sufficient, hydrocarbons will flow to the surface on their pressure alone. In all cases, after a certain amount of production, the pressure decreases enough that punps have to be installed.
- Eventually, the pressure is reduced so that pumps cannot lift the Hydrocarbons. In this case, secondary and tertiary production techniques are started

All of these procedures are designed to keep reservoir pressure high. They generally involve injection various types of fluids into the reservoir. Some of the types of fluids include: gases- *e.g.* hydrocarbon gases from the field or nearby fields, or inert other gases such as nitrogen or carbon dioxide water- this is probably the most common type of secondary recovery, so these type of operations are often called water floods.

The water that is injected has to have a very specific chemistry because water that is different from the formation fluids may adverse effects on the formation- *e.g.* precipitate salts or cause clays to swell.

8

Oil and Gas

OVERVIEW

The Indian oil and gas sector is one of the six core industries in India and has very significant forward linkages with the entire economy.

India has been growing at a decent rate annually and is committed to accelerate the growth momentum in the years to come. This would translate into India's energy needs growing many times in the years to come.

Hence, there is an emphasized need for wider and more intensive exploration for new finds, more efficient and effective recovery, a more rational and optimally balanced global price regime - as against the rather wide upward fluctuations of recent times, and a spirit of equitable common benefit in global energy cooperation. The Indian oil and gas sector is of strategic importance and plays a predominantly pivotal role in influencing decisions in all other spheres of the economy.

The annual growth has been commendable and will accelerate in future consequently encouraging all round growth and development.

This has necessitated the need for a wider intensified search for new fields, evolving better methods of extraction, refining and distribution, the constitution of a national price mechanism - keeping in mind the alarming price fluctuation in the recent past and evolving a spirit of equitable global cooperation.

OIL AND GAS SECTOR

Exploration and Production (E&P)

The growing demand for crude oil and gas in the country and policy initiative of Government of India towards increased E&P activity, have given a great impetus to the Indian E&P industry raising hopes of increased exploration. Oil and Natural Gas Corporation Limited (ONGC) and Oil India Ltd. (OIL), the two National Oil Companies (NOCs) and private and joint-venture companies are engaged in the exploration and production (E&P) of oil and natural gas in the country.

During the year 2008-09, crude oil production has been 33.51 million metric tonnes (MMT) with natural gas at 32.85 billion cubic metre (BCM).Natural gas production is targeted to be about 52.116 BCM.

IMPORTS AND EXPORTS OF CRUDE OIL AND PETROLEUM PRODUCTS

During the financial year 2008-09, imports of crude oil has been 128.16 MMT valued at US$ 73.97 billion. Imports of crude oil during 2007-08 was 121.67 MMT valued at US$ 58.98 billion. This marked an increase of 5.33 per cent during 2008-09 in quantity terms and increased by 25.37 per cent in value terms. During the financial year 2008-09, exports of petroleum products in quantity terms is 36.93 MMT valued at US$ 25.41 billion marking an increase of 6.02 per cent in value terms compared to 2007-08.

New Exploration Licensing Policy (NELP)

New Exploration Licensing Policy (NELP) provides an international class fiscal and contract framework for Exploration and Production of Hydrocarbons. In the first seven rounds of NELP spanning 2000-2009, Production Sharing Contracts (PSCs) for 203 exploration blocks have been signed. Under NELP, 70 oil and gas discoveries have been made by private/joint venture (JV) companies in 20 blocks. With a view to accelerate further the pace of exploration, the eighth round of NELP was launched in April 2009.In the eighth round of NELP,70 exploration blocks comprising of 24 deepwater blocks,28 shallow water blocks and 18 onland blocks will be offered.

Natural Gas

Natural Gas has emerged as one of the most preferred fuel due to its environmentally benign nature, greater efficiency and cost effectiveness. At present, the main producers of natural gas are Oil and Natural Gas Corporation Limited (ONGC), Oil India Limited (OIL) and the Joint Ventures of Panna Mukta and Tapti, and Ravva. Out of the total production of around 96 MMSCMD, after internal consumption, LPG extraction and unavoidable flaring, around 73 MMSCMD is available for sale to various consumers.

In addition, around 7 MMTPA of re-gasified LNG (about 23 MMSCMD) is also being supplied to domestic consumers. Gas produced by ONGC and OIL from the existing nominated blocks is sold at administered prices fixed by the Government. As against a total allocation of 150 MMSCMD of gas, actual supply under APM is presently around 53 MMSCMD.

Oil and Natural Gas Corporation Limited (ONGC)

Oil and Natural Gas Commission (then Commission) wasestablished on 14th August, as a statutory body under Oil and Natural Gas Commission Act (The ONGC Act), for the development of petroleum resources and sale of petroleum products. ONGC was converted into a Public Limited Company under the Companies Act, 1956 and named as "Oil and Natural Gas Corporation Limited" with effect from February.

ONGC Videsh Limited (OVL)

ONGC Videsh Limited (OVL), a wholly owned subsidiary of ONGC, was incorporated as Hydrocarbons India Limited on March 5, 1965 with an initial authorised capital of ₹. 5 Lakhs, for the business of international exploration and production. Its name was changed to ONGC Videsh Limited on June 15, 1989. The authorised and paid-up share capital of OVL as on March

31, 2007 was ₹ 1,000 crore. The primary business of the company is to prospect for oil and gas abroad.

This includes acquisition of oil and gas fields in foreign countries as well as exploration, production, transportation and sale of oil and gas. OVL has presence in 17 countries. It has 37 oil and gas projects. OVL has production of oil and gas from Sudan, Vietnam, Syria, Russia and Colombia. Block BC 10 in Brazil is currently under development with production expected to begin in 2009-10.

Block A-1 and A-3 in Myanmar, North Ramadan Block and NEMED in Egypt, Najwat Najem Structure in Qatar and Farsi Offshore Block in Iran have discoveries and appraisal work is being carried out after which the fields shall be put on development. The remaining projects are in exploration phase. Further, OVL is pursuing acquisition of various oil and gas exploration and production opportunities in Centra Asia, Latin America, Africa, Middle East and South East Asia, which are at different stages.

Oil India Limted (OIL)

Oil India Liimted (OIL), a Government of India Enterprise, under the administrative set-up of Ministry of Petroleum and Natural Gas, is engaged in the business of exploration, production and transportation of crude oil andnatural gas.The authorized capital of the company is ₹500.00 crores and the paid up capital of the company is ₹ 214.00 crore.

OIL produces crude oil and natural gas from its oilfields in Assam and Arunachal Pradesh, non-associated gas from its fields in western Rajasthan and processes LPG from the natural gas in Assam. The Company presently has operational areas in Assam, Arunachal Pradesh, Mizoram, Orissa, Uttar Pradesh, Uttarakhand and Rajasthan in the country.

OIL is operating in 19 nominated ML and 19 nominated PELs. The Company has acquired participating interest in a total of 21 NELP blocks up to the end of NELP-VI bidding round with the right of Operatorship in respect of 12 blocks. The Company also holds Participating Interests (Pis) in another four

Pre-NELP JV blocks in India and Production Sharing Interest (PSI) in one Joint Venture Contract with other partners in Arunachal Pradesh.

OIL is presently active overseas in seven countries, *viz.* Libya, Gabon, Iran, Nigeria, Yemen, Sudan and Bangladesh, pursuing various upstream E&P activities. In addition, the Company is continuously scouting for suitable E&P opportunities in other countries like Syria, Indonesia, Oman, Kazakhastan, Russia, etc., either alone or with suitable partners.

GAIL (India) Limited

GAIL (India) Limited, India's principal Gas Transmission and Marketing Company, was created in 1984 with the objective of accelerating and optimizing the effective and economic use of natural gas and its fractions to the benefit of national economy. In line with core objective of its incorporation, GAIL has, over the years, developed natural gas infrastructure for sustained development of gas market in the country.

GAIL, in the last two decades of its existence, has created a sizeable natural gas market in the country and presently markets around 25 BCM of Natural Gas. GAIL handles around 28 BCM of Natural Gas through its Transmission Network. Currently, GAIL's market share in gas transmission and marketing is 79% and 70% respectively.

REFINING

Refining Capacity

During the year 2008-09, domestic refinery production was 160.77 MMT.By the end of XI plan, refinery capacity is expected to reach 240.96 MMT per annum.The country is net exporter of petroleum products, and products like naphtha, petrol, diesel and Aviation Turbine Fuel (ATF) etc. were also exported during the year. At present, there are 20 refineries operating in the country, out of which 17 are in the public sector and 3 in the private sector. Out of 17 public sector refineries, 8 are owned by Indian Oil Corporation Limited (IOCL), 2 each by Chennai

Petroleum Corporation Limited (a subsidiary of IOCL), Hindustan Petroleum Corporation Limited (HPCL), Bharat Petroleum Corporation Limited (BPCL) and Oil and Natural Gas Corporation Limited, and 1 by Numaligarh Refinery Limited (a subsidiary of BPCL). The private sector refineries belong to Reliance Industries Limited and Essar Oil Limited.

Chennai Petroleum Corporation Limited (CPCL)

Chennai Petroleum Corporation Limited (CPCL) formerly known as Madras Refineries Limited was formed as a joint venture in 1965 between the Government of India (GOI), AMOCO India Inc., U.S.A. and National Iranian Oil Company (NIOC) having a share holding in the ratio 74%: 13%: and 13% respectively. In 1985, AMOCO disinvested in favour of GOI and the shareholding percentage of GOI and NIOC stood revised at 84.62 and 15.38 respectively.

Later, GOI disinvested 16.92% of the paid up capital in favour of Unit Trust of India, Mutual Funds, Insurance Companies and Banks on 19th May 1992, thereby reducing its holding to 67.7%. A public issue of CPCL shares was also made in 1994. As a part of the restructuring steps taken up by the Government of India, Indian Oil Corporation Limited (IOCL) acquired equity from GOI in 2000-01. Currently, IOCL holds 51.88% while Naftiran Inter-trade Company Limited (an affiliate of NIOC) continued its holding at 15.40%.

CPCL has two refineries, with a combined refining capacity of 10.5 million metric tonnes per annum (MMTPA). The Manali Refinery in Chennai has a capacity of 9.5 MMTPA and is one of the most complex refineries in India with Fuel, Lube, Wax and Petrochemical feedstocks production facilities. The second refinery at Cauvery Basin, Nagapattinam was set up initially with a capacity of 0.5 MMTPA in 1993 and later its capacity was enhanced to 1.0 MMTPA in 2002.

Bongaigaon Refinery and Petrochemicals Limited (BRPL)

BRPL was incorporated on February 20,1974, with the objective of installation of Refinery having crude processing capacity of 1 MMTPA and a Petrochemical Complex consisting of Xylene, Dimethyl Terephthalate (DMT) and Polyester Staple Fibre (PSF) units. The crude processing capacity of the Refinery was enhanced to 2.35 MMTPA by commissioning of its Refinery Expansion Units.

The authorized equity capital and the paid-up capital of the Company is ₹200 crore and ₹199.82 crore respectively. The Government of India disinvested its equity share of 74.46% to Indian Oil Corporation Limited (IOCL) in March, 2001 and hence BRPL became the subsidiary Company of IOCL on 29th March, 2001.

Numaligarh Refinery Limited (NRL)

Numaligarh Refinery, Popularly known as "Assam Accord Refinery" had been set up as a grass-root refinery at Numaligarh in the District of Golaghat (Assam) in fulfillment of the commitment made by Government of India in the historic "Assam Accord", signed on 15th august, 1985 for providing the required thrust towards industrial and economic development of Assam.

Both the Refinery and its adjacent Marketing Terminal were completed within the approved project cost of ₹.2724 crore. Commissioning process of Numaligarh Refinery was completed in June 2000 and commercial production commenced from 1st October, 2000.

Mangaore Refinery and Petrochemicals Limited (MRPL)

Mangalore Refinery and Petrochemicals Limited (MRPL), first joint venture company for setting up a crude petroleum Refinery in India was formed in 1987 jointly by Hindustan Petroleum Corporation Limited alongwith Indian Rayon and Industries Limited and its associate companies. The refinery project was commissioned in March, with an actual capacity of 3.69 MMTPA. The expansion project of MRPL, having capacity of 9.69 MMTPA, was commissioned in April.

The refinery is located at Mangalore on the western coast of India, primarily conceived to maximize middle distillates, such as kerosene and diesel. The refinery is designed to process light to heavy and sour to sweet crude. The performance of MRPL started deteriorating after dismantling of APM for refineries in April, 1998 and the Company came very close to becoming a sick company by 2002-03.

With the approval of the Government, ONGCacquired the entire stake of Aditya Birla Group in MRPL for ₹.59.43 crore and also infused additional equity capital of ₹.600 crore in March, 2003 as part of the approved debt restructuring plan. With this, ONGC acquired 51% stake in the equity of MRPL.

In June/July 2003, ONGC acquired 35.80 crore equity shares held by banks and financial institutions issued against part conversion of their loans in terms of debt restructuring plan, increasing its stake in MRPL to 71.62%.

Directorate General of Hydrocarbons (DGH)

The Directorate General of Hydrocarbons (DGH) was established under the administrative control of Ministry of Petroleum and Natural Gas by Government of India Resolution. Objectives of DGH are to promote sound management of the oil and natural gas resources having a balanced regard for environment, safety, technological and economic aspects of the petroleum activity. DGH has been entrusted with certain responsibilities concerning the

Production Sharing Contracts for discovered fields and exploration blocks, promotion of investment and monitoring of E&P activities including review of reservoir performance of major fields.

In addition, DGH is also engaged in opening up of new/ unexplored areas for future exploration and development of nonconventional hydrocarbon energy sources.

Engineers India Limited (EIL)

Engineers India Limited (EIL) was established in 1965 to provide engineering and related technical services for petroleum refineries and other related projects. Over the years, it has diversified into and excelled in various fields. EIL has emerged as Asia's leading design, engineering and turnkey contracting company in Petroleum Refining, Petrochemicals, Chemicals and Fertilizers, Pipelines, Offshore Oil and Gas, Onshore Oil and Gas,Terminals and Storages, Mining and Metallurgy and Infrastructure.Engineers India is an ISO 9001:2000 accredited Company.

Balmer Lawrie and Co. Ltd. (BL)

Balmer Lawrie and Co. Ltd. (BL) was established as a Partnership Firm and was incorporated as Private Limited Company in 1924. It was subsequently converted into a Public Limited Company in the year 1936 with its Registered Office at Kolkata.

Biecco Lawrie Limited (BLL)

Biecco Lawrie Limited (BLL), a Government of India Enterprise, under the administrative control of the Ministry of Petroleum and Natural Gas (MOP&NG), was established in 1919 and became a Government Company in 1972. This is a medium sized Engineering Unit with diversified activities having two factories located at Kolkata.

Oil Industry Development Board (OIDB)

The Oil Industry (Development) Act, 1974 was enacted following successive and steep increase in the international prices of crude oil and petroleum products since early 1973, when the need of progressive self-reliance in petroleum and petroleum based industrial raw materials assumed great importance.

Oil Industry Safety Directorate (OISD)

The Oil Industry Safety Directorate (OISD) assists Safety Council under Ministry of Petroleum and Natural Gas (MOP&NG) headed by Secretary, P&NG as Chairman and includes Additional/Joint Secretaries, Advisors in MOP&NG, Chief Executives of all Public Sector Undertakings (PSUs) under the Ministry, Chief Controller of Explosives (CCE), Advisor

(Fire) of the Govt. of India, DGMS and the Director General of Factory Advice Service and Labour Institute etc. as members.

Centre for High Technlogy (CHT)

Centre for High Technology (CHT) was established in 1987 as a specialized agency of the oil industry to assess futuristic requirements, acquire, develop and adopt technologies in the field of refinery processes, petroleum products, additives, storage and handling of crude oil, products and gas.

Petroleum India International (PII)

Petroleum India International (PII) is a consortium of Public Sector Companies in the petroleum, Petrochemicals and engineering sector. The member companies include Indian Oil Corporation Ltd., Bharat Petroleum Corporation Ltd., Bongaigaon Refinery and Petrochemicals Ltd., Chennai Petroleum Corporation Ltd.,

Engineers India Ltd., Hindustan Petroleum Corporation Ltd, Oil India Ltd and Indian Petrochemicals Corporation Ltd. PII was established in 1986 with the common objectives of mobilizing the individual capabilities of its member companies into a joint endeavour for providing technical managerial and other human resources on a global basis.

Petroleum Planning and Analysis Cell (PPAC)

The Petroleum Planning and Analysis Cell (PPAC) was created w.e.f. 1st April 2002 after dismantling of the Administered Pricing Mechanism (APM) in the petroleum sector and abolition of the erstwhile Oil Coordination Committee (OCC). The Governing Body under the chairmanship of Secretary (PNG) and senior officials of MOPNG and Chief Executives of major oil and gas PSUs as members provides necessary supervision, guidelines in the functioning of PPAC.

INVESTMENT OPPORTUNITIES

- Petroleum products are the single largest merchandise export from India.

- Improved Oil Recovery (IOR)/Enhanced Oil Recovery (EOR) techniques
- Crude oil production from the deepwater block D6 in KG Basin
- Use of improved technology
- Extended oil field acquisition activities
- Capacity utilization of refineries
- Foreign company collaboration
- End-user market and Infrastructure development
- Setting up oil and gas courses at universities and training institutes
- Opportunities for world-class service providers

National Auto Fuel Policy

The Auto Fuel Policy aims to comprehensively and holistically address the issues of vehicular emissions, vehicular technologies, and auto fuel quality in a cost-efficient manner while ensuring the security of fuel supply.

The policy objectives are:

- Ensure sustainable, safe, affordable and uninterrupted supplies of auto fuels of right quality to support social and economic development. One of the key factors for meeting this policy objective is to diversify the sources and reduce dependence on any single source of supply.
- Over the years, infrastructure for the import of crude and crude products,x their processing and production, and storage and transportation has been created in the country. Considerable investment has been made in developing this infrastructure and the logistics for the distribution of petroleum products in the country. The Auto Fuel Policy is committed to an optimal utilization of such an infrastructure.
- Assess the future trends in emission and air quality requirements from the view point of public health, and establishment of a consistent framework within which different policy options to reduce emissions can be

assessed. It is, therefore, required that environmental objectives for air quality be determined, emission reduction targets be established, input data on costs and benefits be collected and cost effective measures to reduce emissions be identified. Appropriate institutional arrangements to be put in place to where such activities can be handled effectively.

- Adopt such vehicular emission standards that they together with other measures, will be able to make a decisive impact on air quality, without placing an undue burden on the people.
- Vehicular emission standards and auto fuel quality should offer choice to the citizens and equally a choice to automobile manufactures in matters of technology selection. Principles of widening the choice and promoting competition amongst automobile technologies, within the limits that are imposed by the availability of auto fuels and security of their supplies.
- As elsewhere in the world, the Government should decide only the vehicular mission standards and the corresponding fuel specifications without specifying vehicle technology and the type of fuel.
- The requirement of investments to reach vehicular technology and fuel quality of Euro III equivalent levels throughout the country is estimated in the range of ₹ 50,000 - ₹ 60,000 crore. Therefore, to achieve the air quality targets by gradually improving emission standards and a phased up gradation of fuel quality and vehicular technology, taking note of the financial, technical and institutional considerations as also the absorptive capacity is required.
- Administered fuel prices, carrying subsidies and cross-subsidies, lead to distortions in fuel usage pattern. Determination of fuel prices on the principles of import parity and putting in place a medium term fiscal regime as early as possible are necessary for the sustainability of fuel usage pattern.

- In order to remain relevant, the Auto Fuel Policy must undergo periodic revisions, preferably at an interval of five years. This will allow adjustments in the Policy that may become necessary on account of the technological and other changes that are inevitable in the country and the world. It would also afford an opportunity to different stakeholders to express their views in the light of the changes that take place with time.

FDI POLICY

The present policy on FDI in the Petroleum and Natural Gas sector vide Press Note No 5 (2008) permits FDI up to 100% under the automatic route in all activities other than refining and including market study and formulation, investment/ financing, setting up infrastructure for marketing in Petroleum and Natural Gas Sector subject to sectoral policy.

In Refining, FDI up to 49% in case of Public Sector Undertakings, without involving any divestment or dilution of domestic equity in existing public sector undertakings through Foreign Investment Promotion Board (FIPB) and FDI up to 100% is permitted in case of Private companies under Automatic route subject to sectoral policy.

KEY PLAYERS

- Indian Oil
- Reliance
- Bharat Petroleum
- HP
- ONGC
- BP
- BG Group.
- Gaz de France
- Chevron

The present policy on FDI in the petroleum and natural gas sector vide press note no 5 permits FDI up to 100% under the automatic route in all activities other than refining, investment/financing, setting up infrastructure for marketing in petroleum and natural gas sector subject to sectoral policy including market study and formulation.

In refining, FDI up to 49% in case of public sector undertakings, without involving any disinvestment or dilution of domestic equity in existing public sector undertakings through foreign investment promotion board (FIPB) and FDI up to 100% is permitted in case of private companies under automatic route subject to sectoral policy.

PETROLEUM FLUIDS

REFINING OF PETROLEUM

Fig. 8.1

Petroleum is a complex mixture of organic liquids called crude oil and natural gas, which occurs naturally in the ground and was formed millions of years ago. Crude oil varies from oilfield to oilfield in colour and composition, from a pale yellow low viscosity liquid to heavy black 'treacle' consistencies.

Crude oil and natural gas are extracted from the ground, on land or under the oceans, by sinking an oil well and are then transported by pipeline and/or ship to refineries where their components are processed into refined products. Crude oil and natural gas are of little use in their raw state; their value lies in what is created from them: fuels, lubricating oils, waxes, asphalt, petrochemicals and pipeline quality natural gas. An oil refinery is an organised and coordinated arrangement of manufacturing processes designed to produce physical and chemical changes in crude oil to convert it into everyday products like petrol, diesel, lubricating oil, fuel oil and bitumen. As crude oil comes from the well it contains a mixture of hydrocarbon compounds and relatively small quantities of other materials such as oxygen, nitrogen, sulphur, salt and water.

In the refinery, most of these non - hydrocarbon substances are removed and the oil is broken down into its various components, and blended into useful products. Natural gas from the well, while principally methane, contains quantities of other hydrocarbons - ethane, propane, butane, pentane and also carbon dioxide and water. These components are separated from the methane at a gas fractionation plant.

PETROLEUM HYDROCARBON STRUCTURES

Petroleum consists of three main hydrocarbon groups:

Paraffins

These consist of straight or branched carbon rings saturated with hydrogen atoms, the simplest of which is methane (CH_4) the main ingredient of natural gas. Others in this group include ethane (C_2H_6), and propane (C_3H_8).

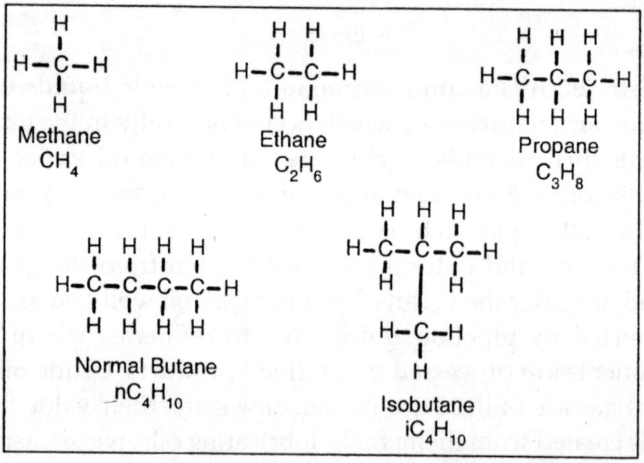

Methane CH_4

Ethane C_2H_6

Propane C_3H_8

Normal Butane nC_4H_{10}

Isobutane iC_4H_{10}

Fig. 8.2

Hydrocarbons

With very few carbon atoms (C_1 to C_4) are light in density and are gases under normal atmospheric pressure. Chemically paraffins are very stable compounds.

Naphthenes

Naphthenes consist of carbon rings, sometimes with side chains, saturated with hydrogen atoms. Naphthenes are chemically stable, they occur naturally in crude oil and have properties similar to paraffins.

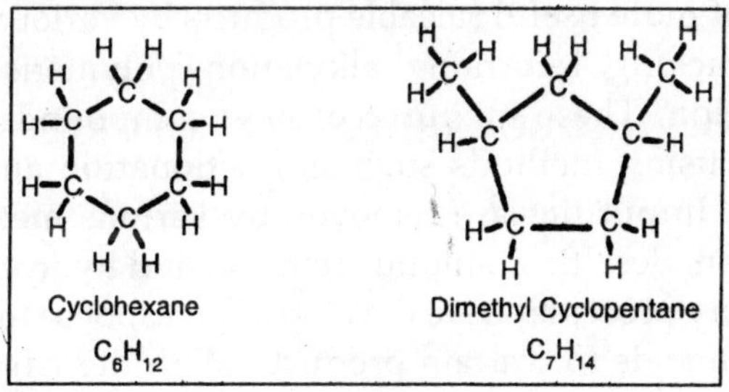

Cyclohexane C_6H_{12}

Dimethyl Cyclopentane C_7H_{14}

Fig. 8.3

Aromatics

Aromatic hydrocarbons are compounds that contain a ring of six carbon atoms with alternating double and single bonds and six attached hydrogen atoms. This type of structure is known as a benzene ring. They occur naturally in crude oil, and can also be created by the refining process.

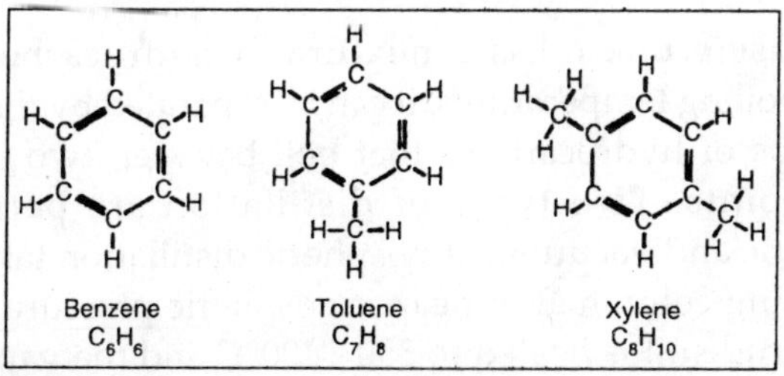

Benzene C_6H_6

Toluene C_7H_8

Xylene C_8H_{10}

Fig. 8.4

The more carbon atoms a hydrocarbon molecule has, the "heavier" it is (the higher is its molecular weight) and the higher is its the boiling point. Small quantities of a crude oil may be composed of compounds containing oxygen, nitrogen, sulphur and metals. Sulphur content ranges from traces to more than 5 per cent. If a crude oil contains appreciable quantities of sulphur it is called a sour crude; if it contains little or no sulphur it is called a sweet crude.

THE REFINING PROCESS

The fractions are further treated to convert them into mixtures of more useful saleable products by various methods such as cracking, reforming, alkylation, polymerisation and isomerisation. These mixtures of new compounds are then separated using methods such as fractionation and solvent extraction. Impurities are removed by various methods, *e.g.* dehydration, desalting, sulphur removal and hydrotreating.

Refinery processes have developed in response to changing market demands for certain products. With the advent of the internal combustion engine the main task of refineries became the production of petrol. The quantities of petrol available from distillation alone was insufficient to satisfy consumer demand. Refineries began to look for ways to produce more and better quality petrol.

Two types of processes have been developed:

1. Breaking down large, heavy hydrocarbon molecules
2. Reshaping or rebuilding hydrocarbon molecules.

DISTILLATION (FRACTIONATION)

Because crude oil is a mixture of hydrocarbons with different boiling temperatures, it can be separated by distillation into groups of hydrocarbons that boil between two specified boiling points. Two types of distillation are performed: atmospheric and vacuum. Atmospheric distillation takes place in a distilling column at or near atmospheric pressure.

The crude oil is heated to 350 - 400°C and the vapour and liquid are piped into the distilling column. The liquid falls to the bottom and the vapour rises, passing through a series of perforated trays (sieve trays). Heavier hydrocarbons condense more quickly and settle on lower trays and lighter hydrocarbons remain as a vapour longer and condense on higher trays.

Liquid fractions are drawn from the trays and removed. In this way the light gases, methane, ethane, propane and butane pass out the top of the column, petrol is formed in the top trays, kerosene and gas oils in the middle, and fuel oils at the bottom. Residue drawn of the bottom may be burned as fuel, processed into lubricating oils, waxes and bitumen or used as feedstock for cracking units.

To recover additional heavy distillates from this residue, it may be piped to a second distillation column where the process is repeated under vacuum, called vacuum distillation.This allows heavy hydrocarbons with boiling points of 450° C and higher to be separated without them partly cracking into unwanted products such as coke and gas. The heavy distillates recovered by vacuum distillation can be converted into lubricating oils by a variety of processes. The most common of these is called solvent extraction.

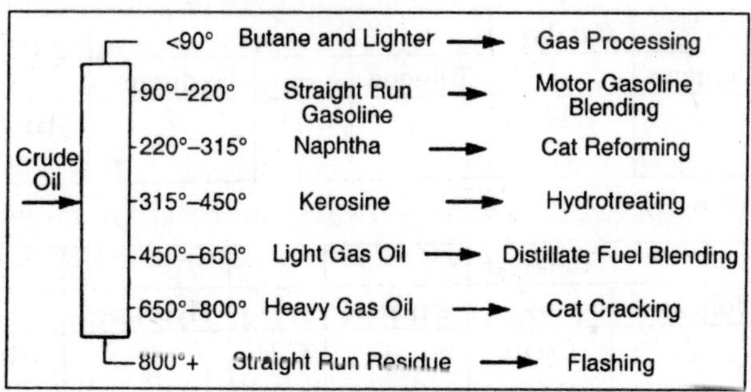

Fig. 8.5 Distilling Crude and Product Disposition

In one version of this process the heavy distillate is washed with a liquid which does not dissolve in it but which dissolves (and so extracts) the non-lubricating oil components out of it. Another version uses a liquid which does not dissolve in it but which causes the non-lubricating oil components to precipitate (as an extract) from it. Other processes exist which remove impurities by adsorption onto a highly porous solid or which remove any waxes that may be present by causing them to crystallise and precipitate out.

REFORMING

Reforming is a process which uses heat, pressure and a catalyst (usually containing platinum) to bring about chemical reactions which upgrade naphthas into high octane petrol and petrochemical feedstock.

The naphthas are hydrocarbon mixtures containing many paraffins and naphthenes. In Australia, this naphtha feedstock comes from the crudes oil distillation or catalytic cracking processes, but overseas it also comes from thermal cracking and hydrocracking processes. Reforming converts a portion of these compounds to isoparaffins and aromatics, which are used to blend higher octane petrol.

- Paraffins are converted to isoparaffins
- Paraffins are converted to naphthenes
- Naphthenes are converted to aromatics

	Catalyst				
Heptane	→	Toluene	+	Hydrogen	
	C_7H_{16}	→	C_7H_8	+	$4H_2$

	Catalyst				
Cyclohexane	→	Benzene	+	Hydrogen	
	C_6H_{12}	→	C_6H_6	+	$3H_2$

CRACKING

Cracking processes break down heavier hydrocarbon molecules (high boiling point oils) into lighter products such as petrol and diesel. These processes include catalytic cracking, thermal cracking and hydrocracking.*e.g.*

A Typical Reaction:

Catalyst				
$C_{16}H_{34}$	→	C_8H_{18}	+	C_8H_{16}

Catalytic cracking is used to convert heavy hydrocarbon fractions obtained by vacuum distillation into a mixture of more useful products such as petrol and light fuel oil. In this process, the feedstock undergoes a chemical breakdown, under controlled heat (450 - 500°C) and pressure, in the presence of a catalyst - a substance which promotes the reaction without itself being chemically changed. Small pellets of silica - alumina or silica - magnesia have proved to be the most effective catalysts.

The cracking reaction yields petrol, LPG, unsaturated olefin compounds, cracked gas oils, a liquid residue called cycle oil, light gases and a solid coke residue. Cycle oil is recycled to cause further breakdown and the coke, which forms a layer on the catalyst, is removed by burning. The other products are passed through a fractionator to be separated and separately processed.

Fluid catalytic cracking uses a catalyst in the form of a very fine powder which flows like a liquid when agitated by steam, air or vapour. Feedstock entering the process immediately meets a stream of very hot catalyst and vaporises. The resulting vapours keep the catalyst fluidised as it passes into the reactor, where the cracking takes place and where it is fluidised by the hydrocarbon vapour.

It then passes to a regenerator vessel where it is fluidised by a mixture of air and the products of combustion which are produced as the coke on the catalyst is burnt off. The catalyst then flows back to the reactor. The catalyst thus undergoes a continuous circulation between the reactor, stripper and regenerator sections. The catalyst is usually a mixture of aluminium oxide and silica. Most recently, the introduction of synthetic zeolite catalysts has allowed much shorter reaction times and improved yields and octane numbers of the cracked gasolines. Thermal cracking uses heat to break down the residue from vacuum distillation. The lighter elements produced from this process can be made into distillate fuels and petrol. Cracked gases are converted to petrol blending components by alkylation or polymerisation. Naphtha is upgraded to high quality petrol by reforming. Gas oil can be used as diesel fuel or can be converted to petrol by hydrocracking. The heavy residue is converted into residual oil or coke which is used in the manufacture of electrodes, graphite and carbides.

Hydrocracking can increase the yield of petrol components, as well as being used to produce light distillates. It produces no residues, only light oils. Hydrocracking is catalytic cracking in the presence of hydrogen. The extra hydrogen saturates, or hydrogenates, the chemical bonds of the cracked hydrocarbons and creates isomers with the desired characteristics. Hydrocracking is also a treating process, because the hydrogen combines with contaminants such as sulphur and nitrogen, allowing them to be removed.

Gas oil feed is mixed with hydrogen, heated, and sent to a reactor vessel with a fixed bed catalyst, where cracking and hydrogenation take place. Products are sent to a fractionator to be separated. The hydrogen is recycled. Residue from this reaction is mixed again with hydrogen, reheated, and sent to a second reactor for further cracking under higher temperatures and pressures.

In addition to cracked naphtha for making petrol, hydrocracking yields light gases useful for refinery fuel, or alkylation as well as components for high quality fuel oils, lube oils and petrochemical feedstocks.

Following the cracking processes it is necessary to build or rearrange some of the lighter hydrocarbon molecules into high quality petrol or jet fuel blending components or into petrochemicals. The former can be achieved by several chemical process such as alkylation and isomerisation.

ALKYLATION

Olefins such as propylene and butylene are produced by catalytic and thermal cracking. Alkylation refers to the chemical bonding of these light molecules with isobutane to form larger branched-chain molecules (isoparaffins) that make high octane petrol. Olefins and isobutane are mixed with an acid catalyst and cooled. They react to form alkylate, plus some normal butane, isobutane and propane. The resulting liquid is neutralised and separated in a series of distillation columns. Isobutane is recycled as feed and butane and propane sold as liquid petroleum gas (LPG). *e.g.*

			Catalyst		
Isobutane	+	Butylene	→	Isooctane	
	C_4H_{10}	+	C_4H_8	→	C_8H_{18}

ISOMERISATION

Isomerisation refers to chemical rearrangement of straight-chain hydrocarbons (paraffins), so that they contain branches attached to the main chain (isoparaffins).

This is done for two reasons:

1. They create extra isobutane feed for alkylation
2. They improve the octane of straight run pentanes and hexanes and hence make them into better petrol blending components.

Isomerisation is achieved by mixing normal butane with a little hydrogen and chloride and allowed to react in the presence of a catalyst to form isobutane, plus a small amount of normal butane and some lighter gases. Products are separated in a fractionator. The lighter gases are used as refinery fuel and the butane recycled as feed. Pentanes and hexanes are the lighter components of petrol. Isomerisation can be used to improve petrol quality by converting these hydrocarbons to higher octane isomers. The process is the same as for butane isomerisation.

POLYMERISATION

Under pressure and temperature, over an acidic catalyst, light unsaturated hydrocarbon molecules react and combine with each other to form larger hydrocarbon molecules. Such process can be used to react butenes (olefin molecules with four carbon atoms) with iso-butane (branched paraffin molecules, or isoparaffins, with four carbon atoms) to obtain a high octane olefinic petrol blending component called polymer gasoline.

HYDROTREATING AND SULPHUR PLANTS

A number of contaminants are found in crude oil. As the fractions travel through the refinery processing units, these impurities can damage the equipment, the catalysts and the quality of the products. There are also legal limits on the contents of some impurities, like sulphur, in products.

Hydrotreating is one way of removing many of the contaminants from many of the intermediate or final products. In the hydrotreating process, the entering feedstock is mixed with hydrogen and heated to 300– 380°C.

The oil combined with the hydrogen then enters a reactor loaded with a catalyst which promotes several reactions:

- Hydrogen combines with sulphur to form hydrogen sulphide (H_2S)
- Nitrogen compounds are converted to ammonia
- Any metals contained in the oil are deposited on the catalyst
- Some of the olefins, aromatics or naphthenes become saturated with hydrogen to become paraffins and some cracking takes place, causing the creation of some methane, ethane, propane and butanes.

SULPHUR RECOVERY PLANTS

The hydrogen sulphide created from hydrotreating is a toxic gas that needs further treatment.

The usual process involves two steps:

1. The removal of the hydrogen sulphide gas from the hydrocarbon stream
2. The conversion of hydrogen sulphide to elemental sulphur, a non-toxic and useful chemical.

Solvent extraction, using a solution of diethanolamine (DEA) dissolved in water, is applied to separate the hydrogen sulphide gas from the process stream. The hydrocarbon gas stream containing the hydrogen sulphide is bubbled through a solution of diethanolamine solution (DEA) under high pressure, such that the hydrogen sulphide gas dissolves in the DEA. The DEA and hydrogen mixture is the heated at a low pressure and the dissolved hydrogen sulphide is released as a concentrated gas stream which is sent to another plant for conversion into sulphur.

Conversion of the concentrated hydrogen sulphide gas into sulphur occurs in two stages:

1. Combustion of part of the H_2S stream in a furnace, producing sulphur dioxide (SO_2) water (H_2O) and sulphur (S).
$$2H_2S + 2O_2 \rightarrow SO_2 + S + 2H_2O$$
2. Reaction of the remainder of the H_2S with the combustion products in the presence of a catalyst. The H_2S reacts with the SO_2 to form sulphur.
$$2H_2S + 2O_2 \rightarrow 3S + 2H_2O$$

As the reaction products are cooled the sulphur drops out of the reaction vessel in a molten state. Sulphur can be stored and shipped in either a molten or solid state.

REFINERIES AND THE ENVIRONMENT

Air, water and land can all be affected by refinery operations. Refineries are well aware of their responsibility to the community and employ a variety of processes to safeguard the environment.

Air

Preserving air quality around a refinery involves controlling the following emissions:

- Sulphur oxides
- Hydrocarbon vapours
- Smoke
- Smells

Sulphur enters the refinery in crude oil feed. Gippsland and most other Australian crude oils have a low sulphur content but other crude's may contain up to 5 per cent sulphur. To deal with this refineries incorporate a sulphur recovery unit which operates on the principles described above.

Many of the products used in a refinery produce hydrocarbon vapours. The escape of vapours to atmosphere are prevented by various means. Floating roofs are installed in tanks to prevent evaporation and so that there is no space for vapour to gather in the tanks.

Where floating roofs cannot be used, the vapours from the tanks are collected in a vapour recovery system and absorbed back into the product stream. In addition, pumps and valves are routinely checked for vapour emissions and repaired if a leakage is found.

Smoke is formed when the burning mixture contains insufficient oxygen or is not sufficiently mixed. Modern furnace control systems prevent this from happening during normal operation. Smells are the most difficult emission to control and the easiest to detect. Refinery smells are generally associated with compounds containing sulphur, where even tiny losses are sufficient to cause a noticeable odour.

Water

Aqueous effluent's consist of cooling water, surface water and process water. The majority of the water discharged from the refinery has been used for cooling the various process streams. The cooling water does not actually come into contact with the process material and so has very little contamination. The cooling water passes through large "interceptors" which separate any oil from minute leaks etc., prior to discharge. The cooling water system at Geelong Refinery is a once-through system with no recirculation.

Rainwater falling on the refinery site must be treated before discharge to ensure no oily material washed off process equipment leaves the refinery. This is done first by passing the water through smaller "plant oil catchers", which each treat rainwater from separate areas on the site, and then all the streams pass to large "interceptors" similar to those used for cooling water. The rainwater from the production areas is further treated in a Dissolved Air Flotation (DAF) unit. This unit cleans the water by using a flocculation agent to collect any remaining particles or oil droplets and floating the resulting flock to the surface with millions of tiny air bubbles.

At the surface the flock is skimmed off and the clean water discharged. Process water has actually come into contact with the process streams and so can contain significant contamination. This water is treated in the "sour water treater" where the contaminants (mostly ammonia and hydrogen sulphide) are removed and then recovered or destroyed in a downstream plant.

The process water, when treated in this way, can be reused in parts of the refinery and discharged through the process area rainwater treatment system and the DAF unit.

Any treated process water that is not reused is discharged as Trade Waste to the sewerage system. This trade waste also includes the effluent from the refinery sewage treatment plant and a portion of treated water from the DAF unit. As most refineries import and export many feed materials and products by ship, the refinery and harbour authorities are prepared for spillage from the ship or pier.

In the event of such a spill, equipment is always on standby at the refinery and it is supported by the facilities of the Australian Marine Oil Spill Centre at Geelong, Victoria.

Land

The refinery safeguards the land environment by ensuring the appropriate disposal of all wastes. Within the refinery, all hydrocarbon wastes are recycled through the refinery slops system. This system consists of a network of collection pipes and a series of dewatering tanks. The recovered hydrocarbon is reprocessed through the distillation units.

Wastes that cannot be reprocessed are either recycled to manufacturers (*e.g.* some spent catalysts can be reprocessed), disposed of in EPA-approved facilities off-site, or chemically treated on-site to form inert materials which can be disposed to land-fill within the refinery. Waste movements within the refinery require a "Process liquid, Sludge and Solid waste disposal permit". Wastes that go off-site must have an EPA "Waste Transport Permit".

Downstream Processing

Fig. 8.6 Oil Refinery in California. Photo: CA Energy Commission

Additional processing follows crude distillation, "downstream" (or closer to the refinery gate and the consumer) of the distillation process. Downstream processing is grouped together in this discussion, but encompasses a variety of highly-complex units designed for very different upgrading processes.

Some change the molecular structure of the input with chemical reactions, some in the presence of a catalyst, and some with thermal reactions.

In general, these processes are designed to take heavy, low-valued feedstock-often itself the output from an earlier process-and change it into lighter, higher-valued output. A catalytic cracker, for instance, uses the gasoil (heavy distillate) output from crude distillation as its feedstock and produces additional finished distillates (heating oil and diesel) and gasoline. Sulfur removal is accomplished in ahydrotreater.

A reforming unit produces higher-octane components for gasoline from lower-octane feedstock that was recovered in the distillation process. A coker uses the heaviest output of distillation-the residue or residuum-to produce a lighter feedstock for further processing, as well aspetroleum coke.

CRUDE OIL QUALITY

The physical characteristics of crude oils differ. Crude oil with a similar mix of physical and chemical characteristics, usually produced from a given reservoir, field, or sometimes even a region, constitutes a crude oil "stream." Most simply, crude oils are classified by their density and sulfur content. A common unit of measurement is API gravity-the American Petroleum Institute's measure of specific gravity of crude oil or condensate in degrees.

It is an arbitrary scale expressing the gravity or density of liquid petroleum products. Less dense (or "lighter") crudes generally have a higher share of light hydrocarbons-higher value products-that can be recovered with simple distillation. The denser ("heavier") crude oils produce a greater share of lower-valued products with simple distillation and require additional processing to produce the desired range of products.

Another premium crude oil, Nigeria's Bonny Light, has a high natural yield of middle distillates. By contrast, almost half of the simple distillation yield from Saudi Arabia's Arabian Light, the historical benchmark crude, is a heavy residue ("residuum") that must be reprocessed or sold at a discount to crude oil. Even West Texas Intermediate and Bonny Light have a yield of about one-third residuum after the simple distillation process.

OTHER REFINERY INPUTS

In addition to crude oil that runs through a simple distillation, a variety of other specialized inputs, usually to downstream units, enhance the refiner's capability to make the desired mix of products. Among these products might be unfinished (partly refined) oil, or imported residual fuel oil used as input to a vacuum distillation unit. The supply pattern for "reformulated gasoline" or RFG, the mandated low-pollution product first required in 1995, includes an important share of blending components that are classified as refinery inputs.

These blending components include oxygenates but consist mainly of products that could be classified as finished gasoline in other jurisdictions or products that require little additional blending to be classified as finished gasoline. While they are counted as "refinery inputs," they are brought to saleable specifications in terminals and blending facilities, not in conventional refineries.

MAJOR REFINERY OUTPUTS

The most important refinery product is motor gasoline, a blend of hydrocarbons with boiling ranges from ambient temperatures to about 400° F. The important qualities for gasoline are octane number (antiknock), volatility (starting and vapour lock), and vapour pressure (environmental control). Additives are often used to enhance performance and provide protection against oxidation and rust formation.

Kerosene is a refined middle-distillate petroleum product that finds considerable use as a jet fuel and around the world in cooking and space heating. When used as a jet fuel, some of the critical qualities are freeze point, flash point, and smoke point. Commercial jet fuel has a boiling range of about 375°-525° F, and military jet fuel 130°-550° F. Kerosene, with less-critical specifications, is used for lighting, heating, solvents, and blending into diesel fuel.

Liquified petroleum gas (LPG) consists principally of propane and butane and is produced for use as fuel and is an intermediate material in the manufacture of petrochemicals. The important specifications for proper performance include vapour pressure and control of contaminants. Distillate fuels such as diesel fuels and domestic heating oils have boiling ranges of about 400°-700° F. The desirable qualities required for distillate fuels include controlled flash and pour points, clean burning, no deposit formation in storage tanks, and a proper diesel fuel cetane rating for good starting and combustion.

Residual fuels are heavier oils, known as No. 5 and No. 6 fuel oils, that remain after the distillate fuel oils and lighter hydrocarbons are distilled away in refinery operations. Many marine vessels, power plants, commercial buildings and industrial facilities use residual fuels or combinations of residual and distillate fuels for heating and processing. The two most critical specifications of residual fuels are viscosity and low sulfur content for environmental control.

A variety of solvents, whose boiling points and hydrocarbon composition are closely controlled, are produced in refineries. These include benzene, toluene, and xylene.

Petrochemicals are products derived from crude oil refining, such as ethylene, propylene, butylene, and isobutylene, that are primarily intended for use as chemical feedstocks in the production of plastics, synthetic fibres, synthetic rubbers, and other products.

Lubricants are produced in special refining processes. Additives such as demulsifiers, antioxidants, and viscosity improvers are blended into the base stocks to provide the characteristics required for motor oils, industrial greases, lubricants, and cutting oils. The most critical quality for lubricating-oil base stock is a high viscosity index, which provides for greater consistency under varying temperatures.

Coke is a residue high in carbon content and low in hydrogen that is the final product of thermal decomposition in the condensation process in cracking. It is almost pure carbon with a variety of uses from electrodes to charcoal briquets.

Asphalt is dark brown-to-black cement-like material obtained by petroleum processing and containing bitumens as the predominant component. It used primarily for road construction and roofing materials, and thus must be inert to most chemicals and weather conditions.

PETROLEUM INDUSTRY

BASIC REFINERY PROCESS – DESCRIPTION AND HISTORY

Petroleum refining has evolved continuously in response to changing consumer demand for better and different products. The original requirement was to produce kerosene as a cheaper and better source of light than whale oil. The development of the internal combustion engine led to the production of gasoline and diesel fuels. The evolution of the airplane created a need first for high-octane aviation gasoline and then for jet fuel, a sophisticated form of the original product, kerosene. Present-day refineries produce a variety of products including many required as feedstocks for the petrochemical industry.

DISTILLATION PROCESSES

The first refinery, opened in 1861, produced kerosene by simple atmospheric distillation. Its by-products included tar and naphtha. It was soon discovered that high-quality lubricating oils could be produced by distilling petroleum under vacuum. However, for the next 30 years kerosene was the product consumers wanted.

Two significant events changed this situation:

1. Invention of the electric light decreased the demand for kerosene, and
2. Invention of the internal combustion engine created a demand for diesel fuel and gasoline (naphtha).

THERMAL CRACKING PROCESSES

With the advent of mass production and World War I, the number of gasoline-powered vehicles increased dramatically and the demand for gasoline grew accordingly. However, distillation processes produced only a certain amount of gasoline from crude oil.

In 1913, the thermal cracking process was developed, which subjected heavy fuels to both pressure and intense heat, physically breaking the large molecules into smaller ones to produce additional gasoline and distillate fuels. Visbreaking, another form of thermal cracking, was developed in the late 1930s to produce more desirable and valuable products.

CATALYTIC PROCESSES

Higher-compression gasoline engines required higher-octane gasoline with better antiknock characteristics. The introduction of catalytic cracking and polymerization processes in the mid- to late 1930s met the demand by providing improved gasoline yields and higher octane numbers. Alkylation, another catalytic process developed in the early 1940s, produced more high-octane aviation gasoline and petrochemical feedstocks for explosives and synthetic rubber.

Subsequently, catalytic isomerization was developed to convert hydrocarbons to produce increased quantities of alkylation feedstocks. Improved catalysts and process methods such as hydrocracking and reforming were developed throughout the 1960s to increase gasoline yields and improve antiknock characteristics. These catalytic processes also produced hydrocarbon molecules with a double bond (alkenes) and formed the basis of the modern petrochemical industry.

TREATMENT PROCESSES

Throughout the history of refining, various treatment methods have been used to remove nonhydrocarbons, impurities, and other constituents that adversely affect the properties of finished products or reduce the efficiency of the conversion processes. Treating can involve chemical reaction and/or physical separation. Typical examples of treating are chemical sweetening, acid treating, clay contacting, caustic washing, hydrotreating, drying, solvent extraction, and solvent dewaxing. Sweetening compounds and acids desulfurize crude oil before processing and treat products during and after processing.

Following the Second World War, various reforming processes improved gasoline quality and yield and produced higher-quality products. Some of these involved the use of catalysts and/or hydrogen to change molecules and remove sulfur. A number of the more commonly used treating and reforming processes are described in this chapter of the manual.

HISTORY OF REFINING

Year Process name Purpose By-products, etc. 1862 Atmospheric distillation Produce kerosene Naphtha, tar, etc. 1870 Vacuum distillation Lubricants (original) Asphalt, residual Cracking feedstocks coker feedstocks. Thermal cracking Increase gasoline Residual, bunker fuel 1916 Sweetening Reduce sulfur andamp; odour Sulfur 1930 Thermal reforming Improve octane number Residual 1932 Hydrogenation Remove sulfur Sulfur 1932 Coking Produce gasoline Coke basestocks 1933 Solvent extraction Improve lubricant Aromatics viscosity index 1935 Solvent dewaxing Improve pour point Waxes 1935 Cat. polymerization Improve gasoline Petrochemical yield & octane feedstocks number 1937 Catalytic cracking Higher octane Petrochemical gasoline feedstocks 1939 Visbreaking Reduce viscosity Increased distillate, tar 1940 Alkylation Increase gasoline High-octane aviation octane & yield gasoline 1940 Isomerization Produce alkylation Naphtha feedstock 1942 Fluid catalytic Increase gasoline Petrochemical cracking yield & octane feedstocks 1950 Deasphalting Increase cracking Asphalt feedstock 1952 Catalytic reforming Convert low-quality Aromatics naphtha 1954 Hydrodesulfurization Remove sulfur Sulfur 1956 Inhibitor sweetening Remove mercaptan Disulfides 1957 Catalytic Convert to molecules Alkylation isomerization with high octane feedstocks number 1960 Hydrocracking Improve quality and Alkylation reduce sulfur feedstocks 1974 Catalytic dewaxing Improve pour point Wax 1975 Residual Increase gasoline Heavy residuals hydrocracking yield from residual

BASICS OF CRUDE OIL

Crude oils are complex mixtures containing many different hydrocarbon compounds that vary in appearance and composition from one oil field to another. Crude oils range in consistency from water to tar-like solids, and in colour from clear to black. An average crude oil contains about 84% carbon, 14% hydrogen, 1-3% sulfur, and less than 1% each of nitrogen, oxygen, metals, and salts. Crude oils are generally classified as paraffinic, naphthenic, or aromatic, based on the predominant proportion of similar hydrocarbon molecules.

Mixed-base crudes have varying amounts of each type of hydrocarbon. Refinery crude base stocks usually consist of mixtures of two or more different crude oils. Relatively simple crude-oil assays are used to classify crude oils as paraffinic, naphthenic, aromatic, or mixed. One assay method is based on distillation, and another method (UOP K factor) is based on gravity and boiling points.

More comprehensive crude assays determine the value of the crude (*i.e.*, its yield and quality of useful products) and processing parameters. Crude oils are usually grouped according to yield structure. Crude oils are also defined in terms of API (American Petroleum Institute) gravity. The higher the API gravity, the lighter the crude. For example, light crude oils have high API gravities and low specific gravities.

Crude oils with low carbon, high hydrogen, and high API gravity are usually rich in paraffins and tend to yield greater proportions of gasoline and light petroleum products; those with high carbon, low hydrogen, and low API gravities are usually rich in aromatics. Crude oils that contain appreciable quantities of hydrogen sulfide or other reactive sulfur compounds are called sour. Those with less sulfur are called sweet. Some exceptions to this rule are West Texas crudes, which are always considered sour regardless of their H(2)S content, and Arabian high-sulfur crudes, which are not considered sour because their sulfur compounds are not highly reactive.

BASICS OF HYDROCARBON CHEMISTRY

Crude oil is a mixture of hydrocarbon molecules, which are organic compounds of carbon and hydrogen atoms that may include from one to 60 carbon atoms. The properties of hydrocarbons depend on the number and arrangement of the carbon and hydrogen atoms in the molecules. The simplest hydrocarbon molecule is one carbon atom linked with four hydrogen atoms: methane.

All other variations of petroleum hydrocarbons evolve from this molecule. Hydrocarbons containing up to four carbon atoms are usually gases; those with five to 19 carbon atoms are usually liquids; and those with 20 or more are solids. The refining process uses chemicals, catalysts, heat, and pressure to separate and combine the basic types of hydrocarbon molecules naturally found in crude oil into groups of similar molecules. The refining process also rearranges their structures and bonding patterns into different hydrocarbon molecules and compounds. Therefore it is the type of hydrocarbon, (paraffinic, naphthenic, or aromatic) rather than its specific chemical compounds that is significant in the refining process.

THREE PRINCIPAL GROUPS OR SERIES OF HYDROCARBON COMPOUNDS THAT OCCUR NATURALLY IN CRUDE OIL

PARAFFINS

The paraffinic series of hydrocarbon compounds found in crude oil have the general formula CnH_{2n+2} and can be either straight chains (normal) or branched chains (isomers) of carbon atoms. The lighter, straight-chain paraffin molecules are found in gases and paraffin waxes. Examples of straight-chain molecules are methane, ethane, propane, and butane (gases containing from one to four carbon atoms), and pentane and hexane (liquids with five to six carbon atoms).

The branched-chain (isomer) paraffins are usually found in heavier fractions of crude oil and have higher octane numbers than normal paraffins. These compounds are saturated hydrocarbons, with all carbon bonds satisfied, that is, the hydrocarbon chain carries the full complement of hydrogen atoms.

AROMATICS

Aromatics are unsaturated ring-type (cyclic) compounds which react readily because they have carbon atoms that are deficient in hydrogen. All aromatics have at least one benzene ring (a single-ring compound characterized by three double bonds alternating with three single bonds between six carbon atoms) as part of their molecular structure. Naphthalenes are fused double-ring aromatic compounds. The most complex aromatics, polynuclears (three or more fused aromatic rings), are found in heavier fractions of crude oil.

NAPHTHENES

Naphthenes are saturated hydrocarbon groupings with the general formula C_nH_{2n}, arranged in the form of closed rings (cyclic) and found in all fractions of crude oil except the very lightest. Single-ring naphthenes (monocycloparaffins) with five and six carbon atoms predominate, with two-ring naphthenes (dicycloparaffins) found in the heavier ends of naphtha.

OTHER HYDROCARBONS

Alkenes

Alkenes are mono-olefins with the general formula C_nH_{2n} and contain only one carbon-carbon double bond in the chain. The simplest alkene is ethylene, with two carbon atoms joined by a double bond and four hydrogen atoms. Olefins are usually formed by thermal and catalytic cracking and rarely occur naturally in unprocessed crude oil.

Dienes and Alkynes

Dienes, also known as diolefins, have two carbon-carbon double bonds. The alkynes, another class of unsaturated hydrocarbons, have a carbon-carbon triple bond within the molecule. Both these series of hydrocarbons have the general formula C_nH_{2n-2}. Diolefins such as 1,2-butadiene and 1,3-butadiene, and alkynes such as acetylene occur in C(5) and lighter fractions from cracking. The olefins, diolefins, and alkynes are said to be unsaturated because they contain less than the amount of hydrogen necessary to saturate all the valences of the carbon atoms. These compounds are more reactive than paraffins or naphthenes and readily combine with other elements such as hydrogen, chlorine, and bromine.

Nonhydrocarbons

Sulfur Compounds

Sulfur may be present in crude oil as hydrogen sulfide H_2S, as compounds (*e.g.*, mercaptans, sulfides, disulfides, thiophenes, etc.), or as elemental sulfur. Each crude oil has different amounts and types of sulfur compounds, but as a rule the proportion, stability, and complexity of the compounds are greater in heavier crude-oil fractions. Hydrogen sulfide is a primary contributor to corrosion in refinery processing units. Other corrosive substances are elemental sulfur and mercaptans. Moreover, the corrosive sulfur compounds have an obnoxious odour.

Pyrophoric iron sulfide results from the corrosive action of sulfur compounds on the iron and steel used in refinery process equipment, piping, and tanks. The combustion of petroleum products containing sulfur compounds produces undesirables such as sulfuric acid and sulfur dioxide. Catalytic hydrotreating processes such as hydrodesulfurization remove sulfur compounds from refinery product streams. Sweetening processes either remove the obnoxious sulfur compounds or convert them to odorless disulfides, as in the case of mercaptans.

Oxygen Compounds

Oxygen compounds such as phenols, ketones, and carboxylic acids occur in crude oils in varying amounts.

Nitrogen Compounds

Nitrogen is found in lighter fractions of crude oil as basic compounds, and more often in heavier fractions of crude oil as nonbasic compounds that may also include trace metals such as copper, vanadium, and/or nickel. Nitrogen oxides can form in process furnaces. The decomposition of nitrogen compounds in catalytic cracking and hydrocracking processes forms ammonia and cyanides that can cause corrosion.

Trace Metals

Metals including nickel, iron, and vanadium are often found in crude oils in small quantities and are removed during the refining process. Burning heavy fuel oils in refinery furnaces and boilers can leave deposits of vanadium oxide and nickel oxide in furnace boxes, ducts, and tubes. It is also desirable to remove trace amounts of arsenic, vanadium, and nickel prior to processing as they can poison certain catalysts.

Salts

Crude oils often contain inorganic salts such as sodium chloride, magnesium chloride, and calcium chloride in suspension or dissolved in entrained water (brine). These salts must be removed or neutralized before processing to prevent catalyst poisoning, equipment corrosion, and fouling. Salt corrosion is caused by the hydrolysis of some metal chlorides to hydrogen chloride (HCl) and the subsequent formation of hydrochloric acid when crude is heated. Hydrogen chloride may also combine with ammonia to form ammonium chloride (NH_4Cl), which causes fouling and corrosion.

Carbon Dioxide

Carbon dioxide may result from the decomposition of bicarbonates present in or added to crude, or from steam used in the distillation process.

MAJOR REFINERY PRODUCTS

Gasoline

The most important refinery product is motor gasoline, a blend of hydrocarbons with boiling ranges from ambient temperatures to about 400 degrees F. The important qualities for gasoline are octane number (antiknock), volatility (starting and vapour lock), and vapour pressure (environmental control). Additives are often used to enhance performance and provide protection against oxidation and rust formation.

Kerosene

Kerosene is a refined middle-distillate petroleum product that finds considerable use as a jet fuel and around the world in cooking and space heating. When used as a jet fuel, some of the critical qualities are freeze point, flash point, and smoke point. Commercial jet fuel has a boiling range of about 375-525 degrees

F, and military jet fuel 130-550 degrees F. Kerosene, with less-critical specifications, is used for lighting, heating, Solvents, and Blending Into Diesel Fuel.

Liquefied Petroleum Gas (LPG)

LPG, which consists principally of propane and butane, is produced for use as fuel and is an intermediate material in the manufacture of petrochemicals. The important specifications for proper performance include vapour pressure and control of contaminants.

Distillate Fuels

Diesel fuels and domestic heating oils have boiling ranges of about 400-700 degrees F. The desirable qualities required for distillate fuels include controlled flash and pour points, clean burning, no deposit formation in storage tanks, and a proper diesel fuel cetane rating for good starting and combustion.

Residual Fuels

Many marine vessels, power plants, commercial buildings and industrial facilities use residual fuels or combinations of residual and distillate fuels for heating and processing. The two most critical specifications of residual fuels are viscosity and low sulfur content for environmental control.

Coke and Asphalt

Coke is almost pure carbon with a variety of uses from electrodes to charcoal briquets. Asphalt, used for roads and roofing materials, must be inert to most chemicals and weather conditions.

Solvents

A variety of products, whose boiling points and hydrocarbon composition are closely controlled, are produced for use as solvents. These include benzene, toluene, and xylene.

Petrochemicals

Many products derived from crude oil refining such as

ethylene, propylene, butylene, and isobutylene are primarily intended for use as petrochemical feedstocks in the production of plastics, synthetic fibres, synthetic rubbers, and other products.

Lubricants

Special refining processes produce lubricating oil base stocks. Additives such as demulsifiers, antioxidants, and viscosity improvers are blended into the base stocks to provide the characteristics required for motor oils, industrial greases, lubricants, and cutting oils. The most critical quality for lubricating-oil base stock is a high viscosity index, which provides for greater consistency under varying temperatures.

COMMON REFINERY CHEMICALS

Leaded Gasoline Additives

Tetraethyl lead (TEL) and tetramethyl lead (TML) are additives formerly used to improve gasoline octane ratings but are no longer in common use except in aviation gasoline.

Oxygenates

Ethyl tertiary butyl ether (ETBE), methyl tertiary butyl ether (MTBE), tertiary amyl methyl ether (TAME), and other oxygenates improve gasoline octane ratings and reduce carbon monoxide emissions.

Caustics

Caustics are added to desalting water to neutralize acids and reduce corrosion. They are also added to desalted crude in order to reduce the amount of corrosive chlorides in the tower overheads. They are used in some refinery treating processes to remove contaminants from hydrocarbon streams.

Sulfuric Acid and Hydrofluoric Acid

Sulfuric acid and hydrofluoric acid are used primarily as catalysts in alkylation processes. Sulfuric acid is also used in some treatment processes.

PETROLEUM REFINING OPERATIONS

Introduction

Petroleum refining begins with the distillation, or fractionation, of crude oils into separate hydrocarbon groups. The resultant products are directly related to the characteristics of the crude processed. Most distillation products are further converted into more usable products by changing the size and structure of the hydrocarbon molecules through cracking, reforming, and other conversion processes as discussed in this chapter. These converted products are then subjected to various treatment and separation processes such as extraction, hydrotreating, and sweetening to remove undesirable constituents and improve product quality. Integrated refineries incorporate fractionation, conversion, treatment, and blending operations and may also include petrochemical processing.

Refining Operations

Petroleum refining processes and operations can be separated into five basic areas:

Fractionation

Fractionation (distillation) is the separation of crude oil in atmospheric and vacuum distillation towers into groups of hydrocarbon compounds of differing boiling-point ranges called fractions or cuts.

CONVERSION

Conversion processes change the size and/or structure of hydrocarbon molecules.

These processes include:

- Decomposition (dividing) by thermal and catalytic cracking
- Unification (combining) through alkylation and polymerization, and
- Alteration (rearranging) with isomerization and catalytic reforming

TREATMENT

Treatment processes are intended to prepare hydrocarbon streams for additional processing and to prepare finished products. Treatment may include the removal or separation of aromatics and naphthenes as well as impurities and undesirable contaminants.

Treatment may involve chemical or physical separation such as dissolving, absorption, or precipitation using a variety and combination of processes including desalting, drying, hydrodesulfurizing, solvent refining, sweetening, solvent extraction, and solvent dewaxing.

FORMULATING AND BLENDING

Formulating and blending is the process of mixing and combining hydrocarbon fractions, additives, and other components to produce finished products with specific performance properties.

OTHER REFINING OPERATIONS

Other refinery operations include light-ends recovery, sour-water stripping, solid waste and wastewater treatment, process-water treatment and cooling, storage, and handling, product movement, hydrogen production, acid and tail-gas treatment, and sulfur recovery.

Auxiliary operations and facilities include steam and power generation; process and fire water systems; flares and relief systems; furnaces and heaters; pumps and valves; supply of steam, air, nitrogen, and other plant gases; alarms and sensors; noise and pollution controls; sampling, testing, and inspecting; and laboratory, control room, maintenance, and administrative facilities.

CRUDE OIL PRETREATMENT (DESALTING)

DESCRIPTION

- Crude oil often contains water, inorganic salts, suspended solids, and water-soluble trace metals. As a first step in the refining process, to reduce corrosion, plugging, and fouling of equipment and to prevent poisoning the catalysts in processing units, these contaminants must be removed by desalting (dehydration).

- The two most typical methods of crude-oil desalting, chemical and electrostatic separation, use hot water as the extraction agent. In chemical desalting, water and chemical surfactant (demulsifiers) are added to the crude, heated so that salts and other impurities dissolve into the water or attach to the water, and then held in a tank where they settle out. Electrical desalting is the application of high-voltage electrostatic charges to concentrate suspended water globules in the bottom of the settling tank. Surfactants are added only when the crude has a large amount of suspended solids. Both methods of desalting are continuous. A third and less-common process involves filtering heated crude using diatomaceous earth.
- The feedstock crude oil is heated to between 150° and 350°F to reduce viscosity and surface tension for easier mixing and separation of the water. The temperature is limited by the vapour pressure of the crude-oil feedstock. In both methods other chemicals may be added. Ammonia is often used to reduce corrosion. Caustic or acid may be added to adjust the pH of the water wash. Wastewater and contaminants are discharged from the bottom of the settling tank to the wastewater treatment facility. The desalted crude is continuously drawn from the top of the settling tanks and sent to the crude distillation (fractionating) tower.

Table. Desalting Process

Feedstock	From	Process	Typical Products	To
Crude	Storage	Treating	Desalted crude	Atmospheric distillation tower
			Waste water	Treatment

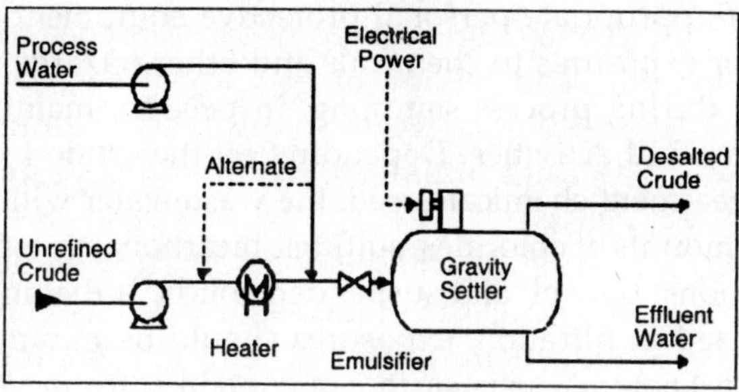

Fig. 8.7 Electrostaitc Desalting.

HEALTH AND SAFETY CONSIDERATIONS

Fire Prevention and Protection

The potential exists for a fire due to a leak or release of crude from heaters in the crude desalting unit. Low boiling point components of crude may also be released if a leak occurs.

Safety

Inadequate desalting can cause fouling of heater tubes and heat exchangers throughout the refinery. Fouling restricts product flow and heat transfer and leads to failures due to increased pressures and temperatures. Corrosion, which occurs due to the presence of hydrogen sulfide, hydrogen chloride, naphthenic (organic) acids, and other contaminants in the crude oil, also causes equipment failure. Neutralized salts (ammonium chlorides and sulfides), when moistened by condensed water, can cause corrosion. Overpressuring the unit is another potential hazard that causes failures.

Health

Because this is a closed process, there is little potential for exposure to crude oil unless a leak or release occurs. Where elevated operating temperatures are used when desalting sour crudes, hydrogen sulfide will be present. There is the possibility of exposure to ammonia, dry chemical demulsifiers, caustics, and/or acids during this operation. Safe work practices and/or the use of appropriate personal protective equipment may be needed for exposures to chemicals and other hazards such as heat, and during process sampling, inspection, maintenance, and turnaround activities. Depending on the crude feedstock and the treatment chemicals used, the wastewater will contain varying amounts of chlorides, sulfides, bicarbonates, ammonia, hydrocarbons, phenol, and suspended solids. If diatomaceous earth is used in filtration, exposures should be minimized or controlled. Diatomaceous earth can contain silica in very fine particle size, making this a potential respiratory hazard.

CRUDE OIL DISTILLATION (FRACTIONATION)

DESCRIPTION

The first step in the refining process is the separation of crude oil into various fractions or straight-run cuts by distillation in atmospheric and vacuum towers. The main fractions or "cuts" obtained have specific boiling-point ranges and can be classified in order of decreasing volatility into gases, light distillates, middle distillates, gas oils, and residuum.

Atmospheric Distillation Tower

- At the refinery, the desalted crude feedstock is preheated using recovered process heat. The feedstock then flows to a direct-fired crude charge heater where it is fed into the vertical distillation column just above the bottom, at pressures slightly above atmospheric and at temperatures ranging from 650° to 700° F (heating crude oil above these temperatures may cause undesirable thermal cracking). All but the heaviest fractions flash into vapour. As the hot vapour rises in the tower, its temperature is reduced. Heavy fuel oil or asphalt residue is taken from the bottom. At successively higher points on the tower, the various major products including lubricating oil, heating oil, kerosene, gasoline, and uncondensed gases (which condense at lower temperatures) are drawn off.

- The fractionating tower, a steel cylinder about 120 feet high, contains horizontal steel trays for separating and collecting the liquids. At each tray, vapors from below enter perforations and bubble caps. They permit the vapors to bubble through the liquid on the tray, causing some condensation at the temperature of that tray. An overflow pipe drains the condensed liquids from each tray back to the tray below, where the higher temperature causes re-evaporation. The evaporation, condensing, and scrubbing operation is repeated many times until the desired degree of product purity is reached. Then side streams from certain trays are taken off to obtain the desired fractions. Products ranging from uncondensed fixed gases at the top to heavy fuel oils at the bottom can be taken continuously from a fractionating tower. Steam is often used in towers to lower the vapour pressure and create a partial vacuum. The distillation process separates the major constituents of crude oil into so-called straight-run products. Sometimes crude oil is "topped" by distilling off only the lighter fractions, leaving a heavy residue that is often distilled further under high vacuum.

Table. Atmospheric Distillation Process

Feedstock	From	Process	Typical Products	To
Crude	Desalting	Separation	Gases	Atmospheric distillation tower
			Naphthas	Reforming or treating
			Kerosene or distillates	Treating
			Gas oil	Catalytic cracking
			Residual	Vacuum tower or visbreaker

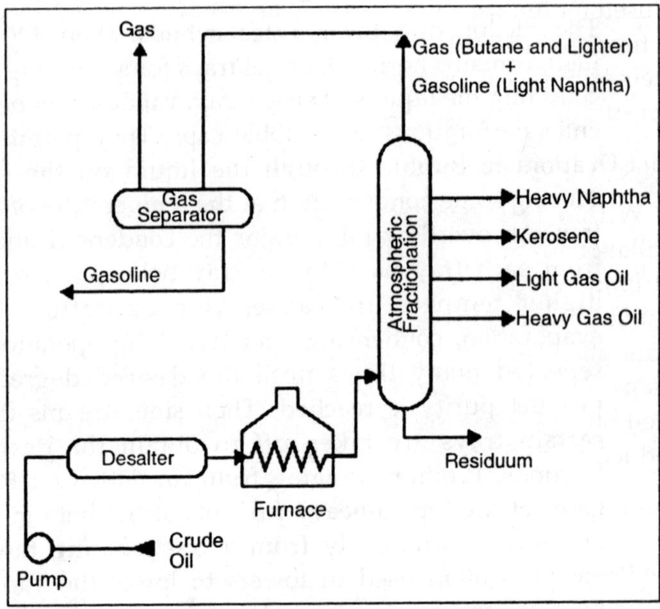

Fig. 8.8 Atmospheric Distillation.

Vacuum Distillation Tower

In order to further distill the residuum or topped crude from the atmospheric tower at higher temperatures, reduced pressure is required to prevent thermal cracking. The process takes place in one or more vacuum distillation towers. The principles of vacuum distillation resemble those of fractional distillation and, except that larger-diameter columns are used to maintain comparable vapour velocities at the reduced pressures, the equipment is also similar.

The internal designs of some vacuum towers are different from atmospheric towers in that random packing and demister pads are used instead of trays. A typical first-phase vacuum tower may produce gas oils, lubricating-oil base stocks, and heavy residual for propane deasphalting.

A second-phase tower operating at lower vacuum may distill surplus residuum from the atmospheric tower, which is not used for lube-stock processing, and surplus residuum from the first vacuum tower not used for deasphalting. Vacuum towers are typically used to separate catalytic cracking feedstock from surplus residuum.

Other Distillation Towers (Columns)

Within refineries there are numerous other, smaller distillation towers called columns, designed to separate specific and unique products. For example, a depropanizer is a small column designed to separate propane and lighter gases from butane and heavier components. Another larger column is used to separate ethyl benzene and xylene. Small "bubble" towers called strippers use steam to remove trace amounts of light products from heavier product streams.

Health and Safety Considerations

Fire Prevention and Protection

Even though these are closed processes, heaters and exchangers in the atmospheric and vacuum distillation units could provide a source of ignition, and the potential for a fire exists should a leak or release occur.

Safety

An excursion in pressure, temperature, or liquid levels may occur if automatic control devices fail. Control of temperature, pressure, and reflux within operating parameters is needed to prevent thermal cracking within the distillation towers. Relief systems should be provided for overpressure and operations monitored to prevent crude from entering the reformer charge.

The sections of the process susceptible to corrosion include (but may not be limited to) preheat exchanger (HCl and H_2S), preheat furnace and bottoms exchanger (H_2S and sulfur compounds), atmospheric tower and vacuum furnace (H_2S, sulfur compounds, and organic acids), vacuum tower (H_2S and organic acids), and overhead (H_2S, HCl, and water). Where sour crudes are processed, severe corrosion can occur in furnace tubing and in both atmospheric and vacuum towers where metal temperatures exceed 450° F. Wet H_2S also will cause cracks in steel.

When processing high-nitrogen crudes, nitrogen oxides can form in the flue gases of furnaces. Nitrogen oxides are corrosive to steel when cooled to low temperatures in the presence of water. Chemicals are used to control corrosion by hydrochloric acid produced in distillation units. Ammonia may be injected into the overhead stream prior to initial condensation and/or an alkaline solution may be carefully injected into the hot crude-oil feed.

If sufficient wash-water is not injected, deposits of ammonium chloride can form and cause serious corrosion. Crude feedstock may contain appreciable amounts of water in suspension which can separate during startup and, along with water remaining in the tower from steam purging, settle in the bottom of the tower.

This water can be heated to the boiling point and create an instantaneous vaporization explosion upon contact with the oil in the unit.

Health

Atmospheric and vacuum distillation are closed processes and exposures are expected to be minimal. When sour (high-sulfur) crudes are processed, there is potential for exposure to hydrogen sulfide in the preheat exchanger and furnace, tower flash zone and overhead system, vacuum furnace and tower, and bottoms exchanger. Hydrogen chloride may be present in the preheat exchanger, tower top zones, and overheads.

Wastewater may contain water-soluble sulfides in high concentrations and other water-soluble compounds such as ammonia, chlorides, phenol, mercaptans, etc., depending upon the crude feedstock and the treatment chemicals.

Safe work practices and/or the use of appropriate personal protective equipment may be needed for exposures to chemicals and other hazards such as heat and noise, and during sampling, inspection, maintenance, and turnaround activities.

Table. Vacuum Distillation Process

Feedstock	From	Process	Typical Products	To
Residuals	Atmospheric tower	Separation	Gas oils	Catalytic cracker
			Lubricants	Hydrotreating or solvent
			Residual	Deasphalter, visbreaker, or coker

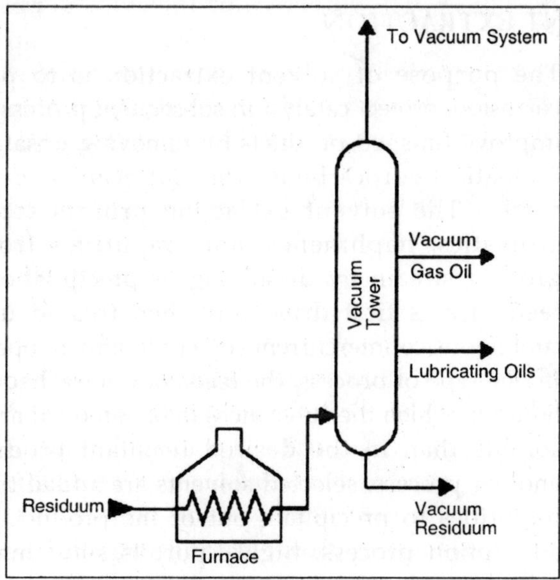

Fig. 8.9 Vacuum Distillation

SOLVENT EXTRACTION AND DEWAXING

DESCRIPTION

Solvent treating is a widely used method of refining. lubricating oils as well as a host of other refinery stocks. Since distillation (fractionation) separates petroleum products into groups only by their boiling-point ranges, impurities may remain. These include organic compounds containing sulfur, nitrogen, and oxygen; inorganic salts and dissolved metals; and soluble salts that were present in the crude feedstock. In addition, kerosene and distillates may have trace amounts of aromatics and naphthenes, and lubricating oil base-stocks may contain wax. Solvent refining processes including solvent extraction and solvent dewaxing usually remove these undesirables at intermediate refining stages or just before sending the product to storage.

SOLVENT EXTRACTION

- The purpose of solvent extraction is to prevent corrosion, protect catalyst in subsequent processes, and improve finished products by removing unsaturated, aromatic hydrocarbons from lubricant and grease stocks. The solvent extraction process separates aromatics, naphthenes, and impurities from the product stream by dissolving or precipitation. The feedstock is first dried and then treated using a continuous countercurrent solvent treatment operation. In one type of process, the feedstock is washed with a liquid in which the substances to be removed are more soluble than in the desired resultant product. In another process, selected solvents are added to cause impurities to precipitate out of the product. In the adsorption process, highly porous solid materials collect liquid molecules on their surfaces.
- The solvent is separated from the product stream by heating, evaporation, or fractionation, and residual trace amounts are subsequently removed from the raffinate by steam stripping or vacuum flashing. Electric precipitation may be used for separation of inorganic compounds. The solvent is then regenerated to be used again in the process.
- The most widely used extraction solvents are phenol,

furfural, and cresylic acid. Other solvents less frequently used are liquid sulfur dioxide, nitrobenzene, and 2,2'-dichloroethyl ether. The selection of specific processes and chemical agents depends on the nature of the feedstock being treated, the contaminants present, and the finished product requirements.

Table. Solvent Extraction Process

Feedstock	From	Process	Typical Products	To
Naphthas, distillates, kerosene	Atm. tower	Treating/ blending	High octane gasoline	Storage
			Refined fuels	Treating and blending
			Spent agents	Treatment and blending

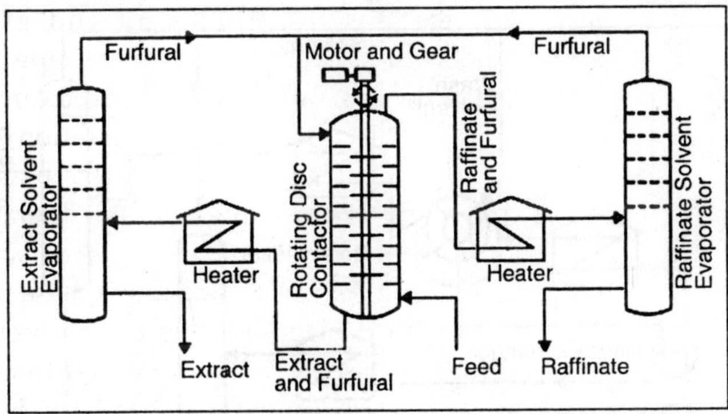

Fig. 8.10 Aromatics Extraction.

There are several processes in use for solvent dewaxing, but all have the same general steps, which are:

1. Mixing the feedstock with a solvent,
2. Precipitating the wax from the mixture by chilling, and
3. Recovering the solvent from the wax and dewaxed oil for recycling by distillation and steam stripping.

Usually two solvents are used: toluene, which dissolves the oil and maintains fluidity at low temperatures, and methyl ethyl ketone (MEK), which dissolves little wax at low temperatures and acts as a wax precipitating agent. Other solvents that are sometimes used include benzene, methyl isobutyl ketone, propane, petroleum naphtha, ethylene dichloride, methylene chloride, and sulfur dioxide. In addition, there is a catalytic process used as an alternate to solvent dewaxing.

Table. Solvent Dewaxing Process

Feedstock	From	Process	Typical Products	To
Lube basestock	Vacuum tower	Treating	Dewaxed lubes	Hydrotreating
			Wax	Hydrotreating
			Spent agents	Treatment or recycle

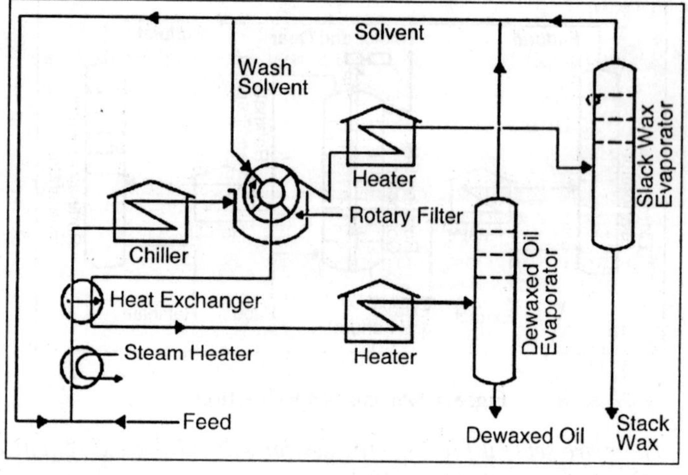

Fig.8.11 Solvent Dewaxing.

HEALTH AND SAFETY CONSIDERATIONS

- *Fire Prevention and Protection*: Solvent treatment is essentially a closed process and, although operating pressures are relatively low, the potential exists for fire from a leak or spill contacting a source of ignition such as the drier or extraction heater. In solvent dewaxing, disruption of the vacuum will create a potential fire hazard by allowing air to enter the unit.

- *Health*: Because solvent extraction is a closed process, exposures are expected to be minimal under normal operating conditions. However, there is a potential for exposure to extraction solvents such as phenol, furfural, glycols, methyl ethyl ketone, amines, and other process chemicals. Safe work practices and/or the use of appropriate personal protective equipment may be needed for exposures to chemicals and other hazards such as noise and heat, and during repair, inspection, maintenance, and turnaround activities.

THERMAL CRACKING

Description

- Because the simple distillation of crude oil produces amounts and types of products that are not consistent with those required by the marketplace, subsequent refinery processes change the product mix by altering the molecular structure of the hydrocarbons. One of the ways of accomplishing this change is through "cracking," a process that breaks or cracks the heavier, higher boiling-point petroleum fractions into more valuable products such as gasoline, fuel oil, and gas oils. The two basic types of cracking are thermal cracking, using heat and pressure, and catalytic cracking.
- The first thermal cracking process was developed around 1913. Distillate fuels and heavy oils were heated under pressure in large drums until they cracked into smaller molecules with better antiknock characteristics. However, this method produced large amounts of solid, unwanted coke. This early process has evolved into the following applications of thermal cracking: visbreaking, steam cracking, and coking.

Visbreaking Process

Visbreaking, a mild form of thermal cracking, significantly lowers the viscosity of heavy crude-oil residue without affecting the boiling point range. Residual from the atmospheric distillation tower is heated (800°-950° F) at atmospheric pressure and mildly cracked in a heater. It is then quenched with cool gas oil to control overcracking, and flashed in a distillation tower.

Table. Visbreaking Process

Feedstock	From	Process	Typical Products	To
Residual	Atmospheric tower & Vacuum tower	Decompose	Gasoline or distillate	Hydrotreating
			Vapor	Hydrotreater
			Residue	Stripper or recycle
			Gases	Gas plant

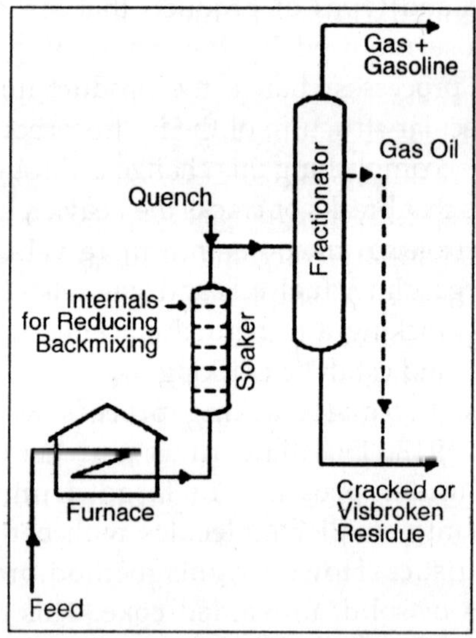

Fig. 8.12 Visbreaking.

Visbreaking is used to reduce the pour point of waxy residues and reduce the viscosity of residues used for blending with lighter fuel oils.

Middle distillates may also be produced, depending on product demand. The thermally cracked residue tar, which accumulates in the bottom of the fractionation tower, is vacuum flashed in a stripper and the distillate recycled.

Steam Cracking Process

Steam cracking is a petrochemical process sometimes used in refineries to produce olefinic raw materials (*e.g.*, ethylene) from various feedstock for petrochemicals manufacture. The feedstock range from ethane to vacuum gas oil, with heavier feeds giving higher yields of by-products such as naphtha.

The most common feeds are ethane, butane, and naphtha. Steam cracking is carried out at temperatures of 1,500°-1,600° F, and at pressures slightly above atmospheric. Naphtha produced from steam cracking contains benzene, which is extracted prior to hydrotreating. Residual from steam cracking is sometimes blended into heavy fuels.

Coking Processes

Coking is a severe method of thermal cracking used to upgrade heavy residuals into lighter products or distillates. Coking produces straight-run gasoline (coker naphtha) and various middle-distillate fractions used as catalytic cracking feedstock. The process so completely reduces hydrogen that the residue is a form of carbon called "coke."

The two most common processes are delayed coking and continuous (contact or fluid) coking. Three typical types of coke are obtained (sponge coke, honeycomb coke, and needle coke) depending upon the reaction mechanism, time, temperature, and the crude feedstock.

Delayed Coking

In delayed coking the heated charge (typically residuum from atmospheric distillation towers) is transferred to large coke drums which provide the long residence time needed to allow the cracking reactions to proceed to completion. Initially the heavy feedstock is fed to a furnace which heats the residuum to high temperatures (900°-950° F) at low pressures (25-30 psi) and is designed and controlled to prevent premature coking in the heater tubes.

The mixture is passed from the heater to one or more coker drums where the hot material is held approximately 24 hours (delayed) at pressures of 25-75 psi, until it cracks into lighter products. Vapors from the drums are returned to a fractionator where gas, naphtha, and gas oils are separated out. The heavier hydrocarbons produced in the fractionator are recycled through the furnace.

After the coke reaches a predetermined level in one drum, the flow is diverted to another drum to maintain continuous operation.

The full drum is steamed to strip out uncracked hydrocarbons, cooled by water injection, and decoked by mechanical or hydraulic methods. The coke is mechanically removed by an auger rising from the bottom of the drum. Hydraulic decoking consists of fracturing the coke bed with high-pressure water ejected from a rotating cutter.

Continuous Coking

Continuous (contact or fluid) coking is a moving-bed process that operates at temperatures higher than delayed coking. In continuous coking, thermal cracking occurs by using heat transferred from hot, recycled coke particles to feedstock in a radial mixer, called a reactor, at a pressure of 50 psi. Gases and vapors are taken from the reactor, quenched to stop any further reaction, and fractionated.

The reacted coke enters a surge drum and is lifted to a feeder and classifier where the larger coke particles are removed as product. The remaining coke is dropped into the preheater for recycling with feedstock. Coking occurs both in the reactor and in the surge drum. The process is automatic in that there is a continuous flow of coke and feedstock.

Table. Coking Processes

Feedstock	From	Process	Typical Products	To
Residual	Atmospheric & vacuum catalytic cracker	Decomposition	Naphtha, gasoline, column, blending	Distillation
Clarified oil	Catalytic cracker		Coke	Shipping, recycle
Tars	Various units		Gas oil	Catalytic cracking
Wasteater (sour)	Treatment			
Gases	Gas plant			

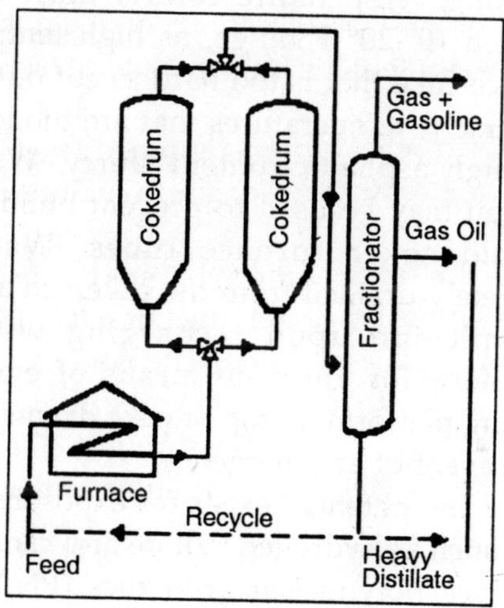

Fig.8.13 Delayed Coking

Health and Safety Considerations

- *Fire Protection and Prevention*: Because thermal cracking is a closed process, the primary potential for fire is from leaks or releases of liquids, gases, or vapors reaching an ignition source such as a heater. The potential for fire is present in coking operations due to vapour or product leaks. Should coking temperatures get out of control, an exothermic reaction could occur within the coker.

- *Safety*: In thermal cracking when sour crudes are processed, corrosion can occur where metal temperatures are between 450° and 900° F. Above 900° F coke forms a protective layer on the metal. The furnace, soaking drums, lower part of the tower, and high-temperature exchangers are usually subject to corrosion. Hydrogen sulfide corrosion in coking can also occur when temperatures are not properly controlled above 900° F. Continuous thermal changes can lead to bulging and cracking of coke drum shells. In coking, temperature control must often be held within a 10°-20° F range, as high temperatures will produce coke that is too hard to cut out of the drum. Conversely, temperatures that are too low will result in a high asphaltic-content slurry. Water or steam injection may be used to prevent buildup of coke in delayed coker furnace tubes. Water must be completely drained from the coker, so as not to cause an explosion upon recharging with hot coke. Provisions for alternate means of egress from the working platform on top of coke drums are important in the event of an emergency.
- *Health*: The potential exists for exposure to hazardous gases such as hydrogen sulfide and carbon monoxide, and trace polynuclear aromatics (PNA's) associated with coking operations. When coke is moved as a slurry, oxygen depletion may occur within confined spaces such as storage silos, since wet carbon will adsorb oxygen. Wastewater may be highly alkaline and contain oil, sulfides, ammonia, and/or phenol. The potential exists in the coking process for exposure to burns when handling hot coke or in the event of a steam-line leak, or from steam, hot water, hot coke, or hot slurry that may be expelled when opening cokers. Safe work practices and/or the use of appropriate personal protective equipment may be needed for exposures to chemicals and other hazards

such as heat and noise, and during process sampling, inspection, maintenance, and turnaround activities. (Note: coke produced from petroleum is a different product from that generated in the steel-industry coking process.)

CATALYTIC CRACKING

- Catalytic cracking breaks complex hydrocarbons into simpler molecules in order to increase the quality and quantity of lighter, more desirable products and decrease the amount of residuals. This process rearranges the molecular structure of hydrocarbon compounds to convert heavy hydrocarbon feedstock into lighter fractions such as kerosene, gasoline, LPG, heating oil, and petrochemical feedstock.
- The three types of catalytic cracking processes are fluid catalytic cracking (FCC), moving-bed catalytic cracking, and Thermofor catalytic cracking (TCC). The catalytic cracking process is very flexible, and operating parameters can be adjusted to meet changing product demand. In addition to cracking, catalytic activities include dehydrogenation, hydrogenation, and isomerization.

Table. Catalytic Cracking Process

Feedstock	From	Process	Typical Products	To
Gas oils	Towers, coker	Decomposition, alteration	Gasoline	Treater or blend
	visbreaker		Gases	Gas plant
			Middle distillates	Hydrotreat, blend, or recycle
Deasphalted oils	Deasphalter		Petrochem feedstock	Petrochem or other
			Residue	Residual fuel blend

FLUID CATALYTIC CRACKING

- The most common process is FCC, in which the oil is cracked in the presence of a finely divided catalyst which is maintained in an aerated or fluidized state by the oil vapors. The fluid cracker consists of a catalyst section and a fractionating section that operate together as an integrated processing unit. The catalyst section contains the reactor and regenerator, which, with the standpipe and riser, forms the catalyst circulation unit. The fluid catalyst is continuously circulated between the reactor and the regenerator using air, oil vapors, and steam as the conveying media.

- A typical FCC process involves mixing a preheated hydrocarbon charge with hot, regenerated catalyst as it enters the riser leading to the reactor. The charge is combined with a recycle stream within the riser, vaporized, and raised to reactor temperature (900°-1,000° F) by the hot catalyst. As the mixture travels up the riser, the charge is cracked at 10-30 psi. In the more modern FCC units, all cracking takes place in the riser. The "reactor" no longer functions as a reactor; it merely serves as a holding vessel for the cyclones. This cracking continues until the oil vapors are separated from the catalyst in the reactor cyclones. The resultant product stream (cracked product) is then charged to a fractionating column where it is separated into fractions, and some of the heavy oil is recycled to the riser.

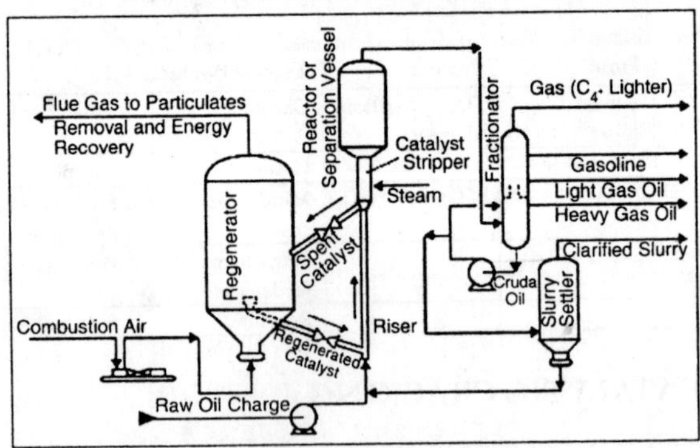

Fig. 8.14 Fluid Catalytic Cracking

- Spent catalyst is regenerated to get rid of coke that collects on the catalyst during the process. Spent catalyst flows through the catalyst stripper to the regenerator, where most of the coke deposits burn off at the bottom where preheated air and spent catalyst are mixed. Fresh catalyst is added and worn-out catalyst removed to optimize the cracking process.

Moving Bed Catalytic Cracking

The moving-bed catalytic cracking process is similar to the FCC process. The catalyst is in the form of pellets that are moved continuously to the top of the unit by conveyor or pneumatic lift tubes to a storage hopper, then flow downward by gravity through the reactor, and finally to a regenerator. The regenerator and hopper are isolated from the reactor by steam seals. The cracked product is separated into recycle gas, oil, clarified oil, distillate, naphtha, and wet gas.

Thermofor Catalytic Cracking

In a typical thermofor catalytic cracking unit, the preheated feedstock flows by gravity through the catalytic reactor bed. The vapors are separated from the catalyst and sent to a fractionating tower. The spent catalyst is regenerated, cooled, and recycled. The flue gas from regeneration is sent to a carbon-monoxide boiler for heat recovery.

Health and Safety Considerations

Fire Prevention and Protection

Liquid hydrocarbons in the catalyst or entering the heated combustion air stream should be controlled to avoid exothermic reactions. Because of the presence of heaters in catalytic cracking units, the possibility exists for fire due to a leak or vapour release. Fire protection including concrete or other insulation on columns and supports, or fixed water spray or fog systems where insulation is not feasible and in areas where firewater hose streams cannot reach, should be considered.

HYDROCRACKING

- Hydrocracking is a two-stage process combining catalytic cracking and hydrogenation, wherein heavier feedstocks are cracked in the presence of hydrogen to produce more desirable products. The process employs high pressure, high temperature, a catalyst, and hydrogen. Hydrocracking is used for feedstocks that are difficult to process by either catalytic cracking or reforming, since these feedstocks are characterized usually by a high polycyclic aromatic content and/or high concentrations of the two principal catalyst poisons, sulfur and nitrogen compounds.
- The hydrocracking process largely depends on the nature of the feedstock and the relative rates of the two competing reactions, hydrogenation and cracking. Heavy aromatic feedstock is converted into lighter products under a wide range of very high pressures (1,000-2,000 psi) and fairly high temperatures (750°-1,500° F), in the presence of hydrogen and special catalysts. When the feedstock has a high paraffinic content, the primary function of hydrogen is to prevent the formation of polycyclic aromatic compounds. Another important role of hydrogen in the hydrocracking process is to reduce tar formation and prevent buildup of coke on the catalyst. Hydrogenation also serves to convert sulfur and nitrogen compounds present in the feedstock to hydrogen sulfide and ammonia.
- Hydrocracking produces relatively large amounts of isobutane for alkylation feedstock. Hydrocracking also performs isomerization for pour-point control and smoke-point control, both of which are important in high-quality jet fuel.

Hydrocracking Process

- In the first stage, preheated feedstock is mixed with recycled hydrogen and sent to the first-stage reactor, where catalysts convert sulfur and nitrogen compounds to hydrogen sulfide and ammonia. Limited hydrocracking also occurs.

- After the hydrocarbon leaves the first stage, it is cooled and liquefied and run through a hydrocarbon separator. The hydrogen is recycled to the feedstock. The liquid is charged to a fractionator. Depending on the products desired (gasoline components, jet fuel, and gas oil), the fractionator is run to cut out some portion of the first stage reactor out-turn. Kerosene-range material can be taken as a separate side-draw product or included in the fractionator bottoms with the gas oil.
- The fractionator bottoms are again mixed with a hydrogen stream and charged to the second stage. Since this material has already been subjected to some hydrogenation, cracking, and reforming in the first stage, the operations of the second stage are more severe (higher temperatures and pressures). Like the outturn of the first stage, the second stage product is separated from the hydrogen and charged to the fractionator.

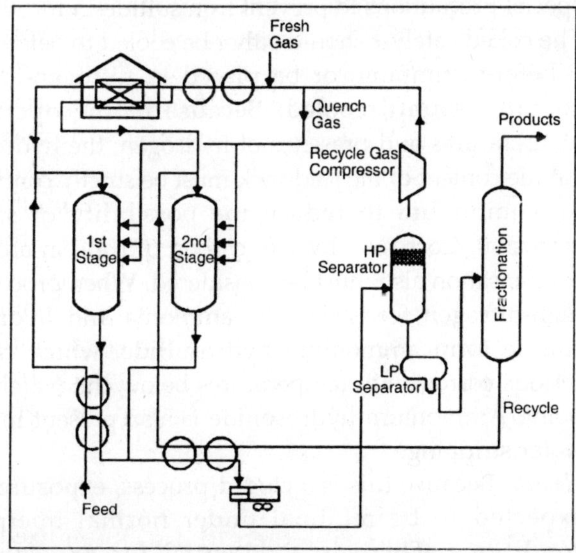

Fig. 8.15 Two-Stage Hydrocracking

Table. Hydrocracking Process.

Feedstock	From	Process	Typical Products	To
High pour point	Catalytic cracker, atmospheric & vacuum tower	Decomposition, hydrogenation	Kerosene, jet fuel	Blending
Gas oil	Vacuum tower, coker		Gasoline, distillates	Blending
Hydrogen	Reformer		Recycle, reformer gas	Gas plant

Health and Safety Considerations

- *Fire Prevention and Protection*: Because this unit operates at very high pressures and temperatures, control of both hydrocarbon leaks and hydrogen releases is important to prevent fires. In some processes, care is needed to ensure that explosive concentrations of catalytic dust do not form during recharging.
- *Safety*: Inspection and testing of safety relief devices are important due to the very high pressures in this unit. Proper process control is needed to protect against plugging reactor beds. Unloading coked catalyst requires special precautions to prevent iron sulfide-induced fires. The coked catalyst should either be cooled to below 120° F before dumping, or be placed in nitrogen-inerted containers until cooled. Because of the operating temperatures and presence of hydrogen, the hydrogen-sulfide content of the feedstock must be strictly controlled to a minimum to reduce the possibility of severe corrosion. Corrosion by wet carbon dioxide in areas of condensation also must be considered. When processing high-nitrogen feedstock, the ammonia and hydrogen sulfide form ammonium hydrosulfide, which causes serious corrosion at temperatures below the water dew point. Ammonium hydrosulfide is also present in sour water stripping.
- *Health*: Because this is a closed process, exposures are expected to be minimal under normal operating conditions. There is a potential for exposure to hydrocarbon gas and vapour emissions, hydrogen and

hydrogen sulfide gas due to high-pressure leaks. Large quantities of carbon monoxide may be released during catalyst regeneration and changeover. Catalyst steam stripping and regeneration create waste streams containing sour water and ammonia. Safe work practices and/or the use of appropriate personal protective equipment may be needed for exposure to chemicals and other hazards such as noise and heat, during process sampling, inspection, maintenance, and turnaround activities, and when handling spent catalyst.

CATALYTIC REFORMING

Description

- Catalytic reforming is an important process used to convert low-octane naphthas into high-octane gasoline blending components called reformates. Reforming represents the total effect of numerous reactions such as cracking, polymerization, dehydrogenation, and isomerization taking place simultaneously. Depending on the properties of the naphtha feedstock (as measured by the paraffin, olefin, naphthene, and aromatic content) and catalysts used, reformates can be produced with very high concentrations of toluene, benzene, xylene, and other aromatics useful in gasoline blending and petrochemical processing. Hydrogen, a significant by-product, is separated from the reformate for recycling and use in other processes.
- There are many different commercial catalytic reforming processes including platforming, powerforming, ultraforming, and Thermofor catalytic reforming. In the platforming process, the first step is preparation of the naphtha feed to remove impurities from the naphtha and reduce catalyst degradation. The naphtha feedstock is then mixed with hydrogen, vaporized, and passed through a series of alternating furnace and fixed-bed reactors containing a platinum

catalyst. The effluent from the last reactor is cooled and sent to a separator to permit removal of the hydrogen-rich gas stream from the top of the separator for recycling. The liquid product from the bottom of the separator is sent to a fractionator called a stabilizer (butanizer). It makes a bottom product called reformate; butanes and lighter go overhead and are sent to the saturated gas plant.

- Some catalytic reformers operate at low pressure (50-200 psi), and others operate at high pressures (up to 1,000 psi). Some catalytic reforming systems continuously regenerate the catalyst in other systems. One reactor at a time is taken off-stream for catalyst regeneration, and some facilities regenerate all of the reactors during turnarounds.

Health and Safety Considerations

- *Fire Prevention and Protection*: This is a closed system; however, the potential for fire exists should a leak or release of reformate gas or hydrogen occur.
- *Safety*: Operating procedures should be developed to ensure control of hot spots during start-up. Safe catalyst handling is very important. Care must be taken not to break or crush the catalyst when loading the beds, as the small fines will plug up the reformer screens. Precautions against dust when regenerating or replacing catalyst should also be considered. Also, water wash should be considered where stabilizer fouling has occurred due to the formation of ammonium chloride and iron salts. Ammonium chloride may form in pretreater exchangers and cause corrosion and fouling. Hydrogen chloride from the hydrogenation of chlorine compounds may form acid or ammonium chloride salt.
- *Health*: Because this is a closed process, exposures are expected to be minimal under normal operating conditions. There is potential for exposure to hydrogen sulfide and benzene should a leak or release occur.

Small emissions of carbon monoxide and hydrogen sulfide may occur during regeneration of catalyst. Safe work practices and/ or appropriate personal protective equipment may be needed for exposures to chemicals and other hazards such as noise and heat during testing, inspecting, maintenance and turnaround activities, and when handling regenerated or spent catalyst.

Table. Catalytic Reforming Process

Feedstock	From	Process	Typical products	To
Desulfurized naphtha	Coker	Rearrange, dehydrogenate	High octane gasoline	Blending
			Aromatics	Petrochemical
Naphthene-rich fractions	hydrocracker, hydrodesulfur		Hydrogen	Recycle, hydrotreat, etc.
Straight-run naphtha	Atmospheric fractionator		Gas	Gas plant

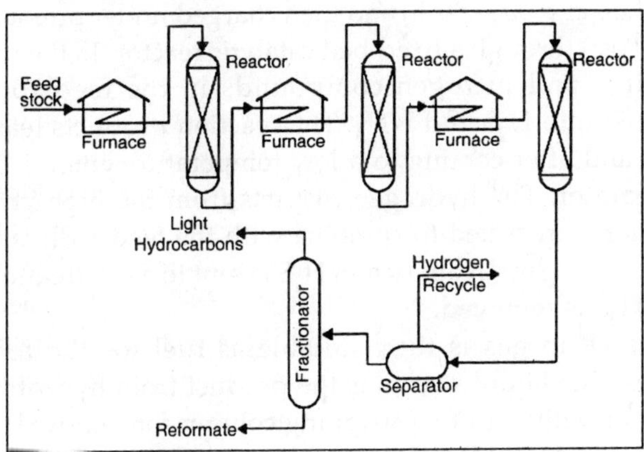

Fig.8.16 Platforming Process.

CATALYTIC HYDROTREATING

Description

Catalytic hydrotreating is a hydrogenation process used to remove about 90% of contaminants such as nitrogen, sulfur, oxygen, and metals from liquid petroleum fractions. These contaminants, if not removed from the petroleum fractions as they travel through the refinery processing units, can have detrimental effects on the equipment, the catalysts, and the quality of the finished product. Typically, hydrotreating is done prior to processes such as catalytic reforming so that the catalyst is not contaminated by untreated feedstock. Hydrotreating is also used prior to catalytic cracking to reduce sulfur and improve product yields, and to upgrade middle-distillate petroleum fractions into finished kerosene, diesel fuel, and heating fuel oils. In addition, hydrotreating converts olefins and aromatics to saturated compounds.

Catalytic Hydrodesulfurization Process

Hydrotreating for sulfur removal is called hydrodesulfurization. In a typical catalytic hydrodesulfurization unit, the feedstock is deaerated and mixed with hydrogen, preheated in a fired heater (600°-800°F) and then charged under pressure (up to 1,000 psi) through a fixed-bed catalytic reactor. In the reactor, the sulfur and nitrogen compounds in the feedstock are converted into H_2S and NH_3. The reaction products leave the reactor and after cooling to a low temperature enter a liquid/gas separator. The hydrogen-rich gas from the high-pressure separation is recycled to combine with the feedstock, and the low-pressure gas stream rich in H_2S is sent to a gas treating unit where H_2S is removed.

The clean gas is then suitable as fuel for the refinery furnaces. The liquid stream is the product from hydrotreating and is normally sent to a stripping column for removal of H_2S and other undesirable components. In cases where steam is used for stripping, the product is sent to a vacuum drier for removal of water. Hydrodesulfurized products are blended or used as catalytic reforming feedstock.

Other Hydrotreating Processes

- Hydrotreating processes differ depending upon the feedstock available and catalysts used. Hydrotreating can be used to improve the burning characteristics of distillates such as kerosene. Hydrotreatment of a kerosene fraction can convert aromatics into naphthenes, which are cleaner-burning compounds.

- Lube-oil hydrotreating uses catalytic treatment of the oil with hydrogen to improve product quality. The objectives in mild lube hydrotreating include saturation of olefins and improvements in colour, odour, and acid nature of the oil. Mild lube hydrotreating also may be used following solvent processing. Operating temperatures are usually below 600° F and operating pressures below 800 psi. Severe lube hydrotreating, at temperatures in the 600°-750° F range and hydrogen pressures up to 3,000 psi, is capable of saturating aromatic rings, along with sulfur and nitrogen removal, to impart specific properties not achieved at mild conditions.
- Hydrotreating also can be employed to improve the quality of pyrolysis gasoline (pygas), a by-product from the manufacture of ethylene. Traditionally, the outlet for pygas has been motor gasoline blending, a suitable route in view of its high octane number. However, only small portions can be blended untreated owing to the unacceptable odour, colour, and gum-forming tendencies of this material. The quality of pygas, which is high in diolefin content, can be satisfactorily improved by hydrotreating, whereby conversion of diolefins into mono-olefins provides an acceptable product for motor gas blending.

Table. Hydrodesulfurization Process

Feedstock	From	Process	Typical Products	To
Naphthas, distillates sour gas oil, residuals	Atmospheric and vacuum tower, catalytic and thermal cracker	Treating, hydrogenation	Naphtha	Blending
			Hydrogen	Recycle
			Distillates	Blending
			H_2S, ammonia	Sulfure plant, treater
			Gas	Gas plant

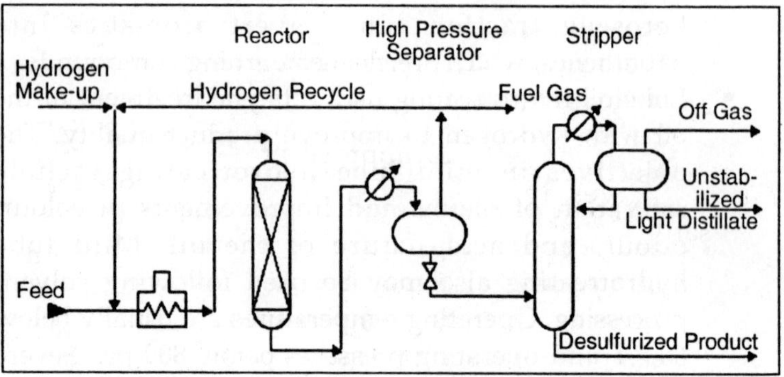

Fig. 8.17 Distillate Hydrodesulfurization.

Health and Safety Considerations

- *Fire Prevention and Protection*: The potential exists for fire in the event of a leak or release of product or hydrogen gas.
- *Safety*: Many processes require hydrogen generation to provide for a continuous supply. Because of the operating temperatures and presence of hydrogen, the hydrogen sulfide content of the feedstock must be strictly controlled to a minimum to reduce corrosion. Hydrogen chloride may form and condense as hydrochloric acid in the lower-temperature parts of the unit. Ammonium hydrosulfide may form in high-temperature, high-pressure units. Excessive contact time and/or temperature will create coking. Precautions need to be taken when unloading coked catalyst from the unit to prevent iron sulfide fires. The coked catalyst should be cooled to below 120° F before removal, or dumped into nitrogen-inerted bins where it can be cooled before further handling. Special antifoam additives may be used to prevent catalyst poisoning from silicone carryover in the coker feedstock.
- *Health*: Because this is a closed process, exposures are

expected to be minimal under normal operating conditions. There is a potential for exposure to hydrogen sulfide or hydrogen gas in the event of a release, or to ammonia should a sour-water leak or spill occur. Phenol also may be present if high boiling-point feedstocks are processed. Safe work practices and/or appropriate personal protective equipment may be needed for exposures to chemicals and other hazards such as noise and heat; during process sampling, inspection, maintenance, and turnaround activities; and when handling amine or exposed to catalyst.

ISOMERIZATION

Description

- Isomerization converts n-butane, n-pentane and n-hexane into their respective isoparaffins of substantially higher octane number. The straight-chain paraffins are converted to their branched-chain counterparts whose component atoms are the same but are arranged in a different geometric structure. Isomerization is important for the conversion of n-butane into isobutane, to provide additional feedstock for alkylation units, and the conversion of normal pentanes and hexanes into higher branched isomers for gasoline blending. Isomerization is similar to catalytic reforming in that the hydrocarbon molecules are rearranged, but unlike catalytic reforming, isomerization just converts normal paraffins to isoparaffins.
- There are two distinct isomerization processes, butane (C_4) and pentane/hexane (C_5/C_6). Butane isomerization produces feedstock for alkylation. Aluminum chloride catalyst plus hydrogen chloride are universally used for the low-temperature processes. Platinum or another metal catalyst is used for the higher-temperature

processes. In a typical low-temperature process, the feed to the isomerization plant is n-butane or mixed butanes mixed with hydrogen (to inhibit olefin formation) and passed to the reactor at 230°-340° F and 200-300 psi. Hydrogen is flashed off in a high-pressure separator and the hydrogen chloride removed in a stripper column. The resultant butane mixture is sent to a fractionator (deisobutanizer) to separate n-butane from the isobutane product.

- Pentane/hexane isomerization increases the octane number of the light gasoline components n-pentane and n-hexane, which are found in abundance in straight-run gasoline. In a typical C_5/C_6 isomerization process, dried and desulfurized feedstock is mixed with a small amount of organic chloride and recycled hydrogen, and then heated to reactor temperature. It is then passed over supported-metal catalyst in the first reactor where benzene and olefins are hydrogenated. The feed next goes to the isomerization reactor where the paraffins are catalytically isomerized to isoparaffins. The reactor effluent is then cooled and subsequently separated in the product separator into two streams: a liquid product (isomerate) and a recycle hydrogen-gas stream. The isomerate is washed (caustic and water), acid stripped, and stabilized before going to storage.

Table. Isomerization Processes

Feedstock	From	Process	Typical Products	To
n-Butane	Various Processes	Rearrangement	Isobutane	Alkylation
n-Pentane			Isopentane	Blending
n-Hexane			Isohexane	Blending
			Gas	Gas Plant

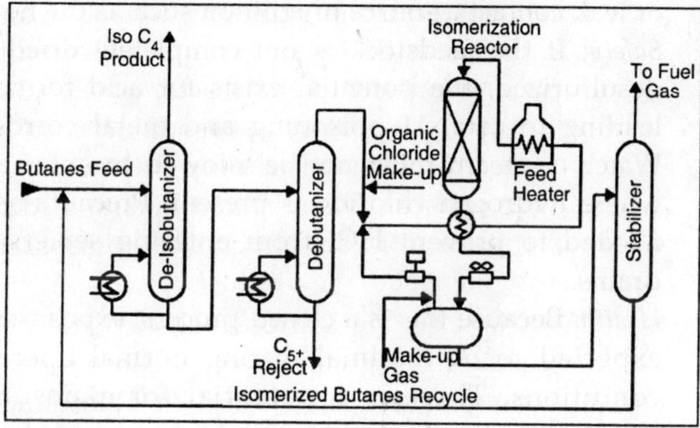

Fig. 8.18 C4 Isomerization.

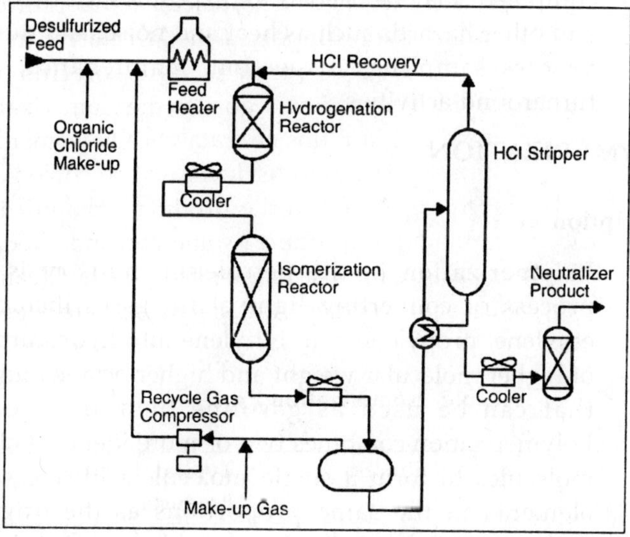

Fig. 8.19 C5 And C6 Isomerization.

Health and Safety Considerations

- *Fire Protection and Prevention*: Although this is a closed process, the potential for a fire exists should a release or leak contact a source of ignition such as the heater.

- *Safety*: If the feedstock is not completely dried and desulfurized, the potential exists for acid formation leading to catalyst poisoning and metal corrosion. Water or steam must not be allowed to enter areas where hydrogen chloride is present. Precautions are needed to prevent HCl from entering sewers and drains.
- *Health*: Because this is a closed process, exposures are expected to be minimal during normal operating conditions. There is a potential for exposure to hydrogen gas, hydrochloric acid, and hydrogen chloride and to dust when solid catalyst is used. Safe work practices and/or appropriate personal protective equipment may be needed for exposures to chemicals and other hazards such as heat and noise, and during process sampling, inspection, maintenance, and turnaround activities.

POLYMERIZATION

Description

- Polymerization in the petroleum industry is the process of converting light olefin gases including ethylene, propylene, and butylene into hydrocarbons of higher molecular weight and higher octane number that can be used as gasoline blending stocks. Polymerization combines two or more identical olefin molecules to form a single molecule with the same elements in the same proportions as the original molecules. Polymerization may be accomplished thermally or in the presence of a catalyst at lower temperatures.
- The olefin feedstock is pretreated to remove sulfur and other undesirable compounds. In the catalytic process the feedstock is either passed over a solid phosphoric acid catalyst or comes in contact with liquid

phosphoric acid, where an exothermic polymeric reaction occurs. This reaction requires cooling water and the injection of cold feedstock into the reactor to control temperatures between 300° and 450° F at pressures from 200 psi to 1,200 psi. The reaction products leaving the reactor are sent to stabilization and/or fractionator systems to separate saturated and unreacted gases from the polymer gasoline product.

Note: In the petroleum industry, polymerization is used to indicate the production of gasoline components, hence the term "polymer" gasoline. Furthermore, it is not essential that only one type of monomer be involved. If unlike olefin molecules are combined, the process is referred to as "copolymerization." Polymerization in the true sense of the word is normally prevented, and all attempts are made to terminate the reaction at the dimer or trimer (three monomers joined together) stage. However, in the petrochemical section of a refinery, polymerization, which results in the production of, for instance, polyethylene, is allowed to proceed until materials of the required high molecular weight have been produced.

HEALTH AND SAFETY CONSIDERATIONS

- *Fire Prevention and Protection*: Polymerization is a closed process where the potential for a fire exists due to leaks or releases reaching a source of ignition.
- *Safety*: The potential for an uncontrolled exothermic reaction exists should loss of cooling water occur. Severe corrosion leading to equipment failure will occur should water make contact with the phosphoric acid, such as during water washing at shutdowns. Corrosion may also occur in piping manifolds, reboilers, exchangers, and other locations where acid may settle out.
- *Health*: Because this is a closed system, exposures are expected to be minimal under normal operating conditions. There is a potential for exposure to caustic wash (sodium hydroxide), to phosphoric acid used in the process or washed out during turnarounds, and

to catalyst dust. Safe work practices and/or appropriate personal protective equipment may be needed for exposures to chemicals and other hazards such as noise and heat, and during process sampling, inspection, maintenance, and turnaround activities.

Table. Polymerization Process

Feedstock	From	Process	Typical Products	To
Olefins	Cracking processes	Unification	High octane naphtha	Gasoline blending
			Petrochem. feedstock	Petrochemical
			Liquefied petro. gas	Storage

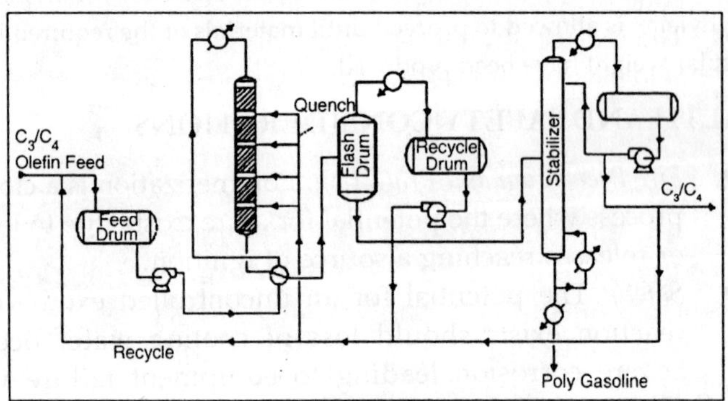

Fig. 8.20 Polymerization Process

ALKYLATION

Description

Alkylation combines low-molecular-weight olefins (primarily a mixture of propylene and butylene) with isobutene in the presence of a catalyst, either sulfuric acid or hydrofluoric acid. The product is called alkylate and is composed of a mixture of high-octane, branched-chain paraffinic hydrocarbons. Alkylate is a premium blending stock because it has exceptional antiknock properties and is clean burning. The octane number of the alkylate depends mainly upon the kind of olefins used and upon operating conditions.

Sulfuric Acid Alkylation Process

- In cascade type sulfuric acid (H_2SO_4) alkylation units, the feedstock (propylene, butylene, amylene, and fresh isobutane) enters the reactor and contacts the concentrated sulfuric acid catalyst (in concentrations of 85% to 95% for good operation and to minimize corrosion). The reactor is divided into zones, with olefins fed through distributors to each zone, and the sulfuric acid and isobutanes flowing over baffles from zone to zone.
- The reactor effluent is separated into hydrocarbon and acid phases in a settler, and the acid is returned to the reactor. The hydrocarbon phase is hot-water washed with caustic for pH control before being successively depropanized, deisobutanized, and debutanized. The alkylate obtained from the deisobutanizer can then go directly to motor-fuel blending or be rerun to produce aviation-grade blending stock. The isobutane is recycled to the feed.

Hydrofluoric Acid Alylation Process

Phillips and UOP are the two common types of hydrofluoric acid alkylation processes in use. In the Phillips process, olefin and isobutane feedstock are dried and fed to a combination reactor/settler system. Upon leaving the reaction zone, the reactor effluent flows to a settler (separating vessel) where the acid separates from the hydrocarbons. The acid layer at the bottom of the separating vessel is recycled. The top layer of hydrocarbons (hydrocarbon phase), consisting of propane, normal butane, alkylate, and excess (recycle) isobutane, is charged to the main fractionator, the bottom product of which is motor alkylate. The main fractionator overhead, consisting mainly of propane, isobutane, and HF, goes to a depropanizer. Propane with trace amount of HF goes to an HF stripper for HF removal and is then catalytically defluorinated, treated, and sent to storage.

Isobutane is withdrawn from the main fractionator and recycled to the reactor/settler, and alkylate from the bottom of the main fractionator is sent to product blending. The UOP process uses two reactors with separate settlers. Half of the dried feedstock is charged to the first reactor, along with recycle and makeup isobutane. The reactor effluent then goes to its settler, where the acid is recycled and the hydrocarbon charged to the second reactor.

Table. Alkylation Process

Feedstock	From	Process	Typical Products	To
Petroleum gas	Distillation or cracking	Unification	High octane gasoline	Blending
Olefins	Cat. or hydro cracking		n-Butane & propane	Stripper or blender
Isobutane	Isomerization			

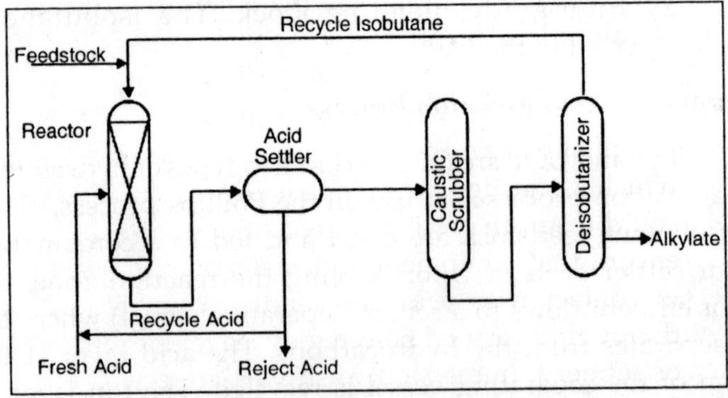

Fig.8.21 Sulfuric Acid Alkylation

The other half of the feedstock also goes to the second reactor, with the settler acid being recycled and the hydrocarbons charged to the main fractionator. Subsequent processing is similar to the Phillips process. Overhead from the main fractionator goes to a depropanizer. Isobutane is recycled to the reaction zone and alkylate is sent to product blending.

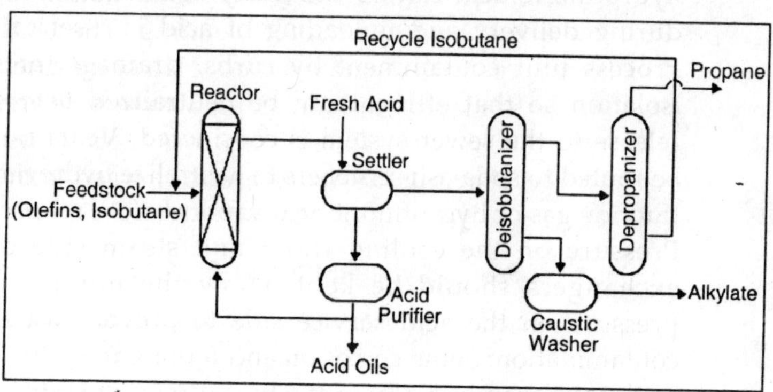

Fig.8.22 Hydrogen Fluoride Alkylation.

Health and Safety Considerations

- *Fire Protection and Prevention*: Alkylation units are closed processes; however, the potential exists for fire should a leak or release occur that allows product or vapour to reach a source of ignition.
- *Safet*: Sulfuric acid and hydrofluoric acid are potentially hazardous chemicals. Loss of coolant water, which is needed to maintain process temperatures, could result in an upset. Precautions are necessary to ensure that equipment and materials that have been in contact with acid are handled carefully and are thoroughly cleaned before they leave the process area or refinery. Immersion wash vats are often provided for neutralization of equipment that has come into contact with hydrofluoric acid. Hydrofluoric acid units

should be thoroughly drained and chemically cleaned prior to turnarounds and entry to remove all traces of iron fluoride and hydrofluoric acid. Following shutdown, where water has been used the unit should be thoroughly dried before hydrofluoric acid is introduced. Leaks, spills, or releases involving hydrofluoric acid or hydrocarbons containing hydrofluoric acid can be extremely hazardous. Care during delivery and unloading of acid is essential. Process unit containment by curbs, drainage, and isolation so that effluent can be neutralized before release to the sewer system is considered. Vents can be routed to soda-ash scrubbers to neutralize hydrogen fluoride gas or hydrofluoric acid vapors before release. Pressure on the cooling water and steam side of exchangers should be kept below the minimum pressure on the acid service side to prevent water contamination. Some corrosion and fouling in sulfuric acid units may occur from the breakdown of sulfuric acid esters or where caustic is added for neutralization. These esters can be removed by fresh acid treating and hot-water washing. To prevent corrosion from hydrofluoric acid, the acid concentration inside the process unit should be maintained above 65% and moisture below 4%.

- *Health*: Because this is a closed process, exposures are expected to be minimal during normal operations. There is a potential for exposure should leaks, spills, or releases occur. Sulfuric acid and (particularly) hydrofluoric acid are potentially hazardous chemicals. Special precautionary emergency preparedness measures and protection appropriate to the potential hazard and areas possibly affected need to be provided. Safe work practices and appropriate skin and respiratory personal protective equipment are needed for potential exposures to hydrofluoric and sulfuric acids during normal operations such as reading gauges, inspecting, and process sampling, as well as during emergency response, maintenance, and turnaround activities. Procedures should be in place to ensure that protective equipment and clothing worn in hydrofluoric acid activities are decontaminated and inspected before reissue. Appropriate personal protection for exposure to heat and noise also may be required.

SWEETENING AND TREATING PROCESSES

Description

- Treating is a means by which contaminants such as organic compounds containing sulfur, nitrogen, and oxygen; dissolved metals and inorganic salts; and soluble salts dissolved in emulsified water are removed from petroleum fractions or streams. Petroleum refiners have a choice of several different treating processes, but the primary purpose of the majority of them is the elimination of unwanted sulfur compounds. A variety of intermediate and finished products, including middle distillates, gasoline, kerosene, jet fuel, and sour gases are dried and sweetened. Sweetening, a major refinery treatment of gasoline, treats sulfur compounds (hydrogen sulfide, thiophene and mercaptan) to improve colour, odour, and oxidation stability. Sweetening also reduces concentrations of carbon dioxide.
- Treating can be accomplished at an intermediate stage in the refining process, or just before sending the finished product to storage. Choices of a treating method depend on the nature of the petroleum fractions, amount and type of impurities in the fractions to be treated, the extent to which the process removes the impurities, and end-product specifications. Treating materials include acids, solvents, alkalis, oxidizing, and adsorption agents.

Acid, Caustic, or Clay Treating

Sulfuric acid is the most commonly used acid treating process. Sulfuric acid treating results in partial or complete removal of unsaturated hydrocarbons, sulfur, nitrogen, and oxygen compounds, and resinous and asphaltic compounds. It is used to improve the odour, colour, stability, carbon residue, and other properties of the oil.

Clay/lime treatment of acid-refined oil removes traces of asphaltic materials and other compounds improving product colour, odour, and stability.

Caustic treating with sodium (or potassium) hydroxide is used to improve odour and colour by removing organic acids (naphthenic acids, phenols) and sulfur compounds (mercaptans, H_2S) by a caustic wash.

By combining caustic soda solution with various solubility promoters (*e.g.*, methyl alcohol and cresols), up to 99% of all mercaptans as well as oxygen and nitrogen compounds can be dissolved from petroleum fractions.

Drying and Sweetening

Feedstocks from various refinery units are sent to gas treating plants where butanes and butenes are removed for use as alkylation feedstock, heavier components are sent to gasoline blending, propane is recovered for LPG, and propylene is removed for use in petrochemicals. Some mercaptans are removed by water-soluble chemicals that react with the mercaptans.

Table. Sweetening and Treating Processes.

Feedstock	From	Process	Typical products	To
Gases, finished products, intermediates	Various	Treatment	Butane & butene	Alkylation
			Propane, distillates	Storage
			Gasoline	Blending
			Propylene	Petrochemical

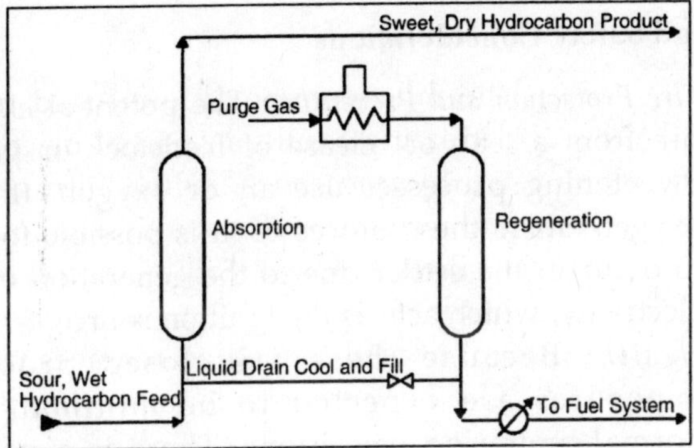

Fig. 8.23 Molecular Sieve Drying and Sweetening.

Caustic liquid (sodium hydroxide), amine compounds (diethanolamine) or fixed-bed catalyst sweetening also may be used. Drying is accomplished by the use of water absorption or adsorption agents to remove water from the products. Some processes simultaneously dry and sweeten by adsorption on molecular sieves.

Sulfur Recovery

Sulfur recovery converts hydrogen sulfide in sour gases and hydrocarbon streams to elemental sulfur. The most widely used recovery system is the Claus process, which uses both thermal and catalytic-conversion reactions. A typical process produces elemental sulfur by burning hydrogen sulfide under controlled conditions. Knockout pots are used to remove water and hydrocarbons from feed gas streams. The gases are then exposed to a catalyst to recover additional sulfur. Sulfur vapour from burning and conversion is condensed and recovered.

Hydrogen Sulfide Scrubbing

Hydrogen sulfide scrubbing is a common treating process in which the hydrocarbon feedstock is first scrubbed to prevent catalyst poisoning. Depending on the feedstock and the nature of contaminants, desulfurization methods vary from ambient temperature-activated charcoal absorption to high-temperature catalytic hydrogenation followed by zinc oxide treating.

Health and Safety Considerations

- *Fire Protection and Prevention*: The potential exists for fire from a leak or release of feedstock or product. Sweetening processes use air or oxygen. If excess oxygen enters these processes, it is possible for a fire to occur in the settler due to the generation of static electricity, which acts as the ignition source.
- *Health*: Because these are closed processes, exposures are expected to be minimal under normal operating conditions. There is a potential for exposure to hydrogen sulfide, caustic (sodium hydroxide), spent caustic, spent catalyst (Merox), catalyst dust and sweetening agents (sodium carbonate and sodium bicarbonate). Safe work practices and/or appropriate personal protective equipment may be needed for exposures to chemicals and other hazards such as noise and heat, and during process sampling, inspection, maintenance, and turnaround activities.

UNSATURATED GAS PLANTS

Description

Unsaturated (unsat) gas plants recover light hydrocarbons (C3 and C4 olefins) from wet gas streams from the FCC, TCC, and delayed coker overhead accumulators or fractionation receivers.

In a typical unsat gas plant, the gases are compressed and treated with amine to remove hydrogen sulfide either before or after they are sent to a fractionating absorber where they are mixed into a concurrent flow of debutanized gasoline.

The light fractions are separated by heat in a reboiler, the offgas is sent to a sponge absorber, and the bottoms are sent to a debutanizer. A portion of the debutanized hydrocarbon is recycled, with the balance sent to the splitter for separation. The overhead gases go to a depropanizer for use as alkylation unit feedstock.

Table. Unsat Gas Plant Process.

Feedstock	From	Process	Typical Products	To
Gas Oils	FCC, TCC, delayed coker	Treatment	Gasoline	Recycle or treating
			Gases	Alkylation

Health and Safety Considerations

- *Fire Prevention and Protection*: The potential of a fire exists should spills, releases, or vapors reach a source of ignition.
- *Safety*: In unsat gas plants handling FCC feedstock, the potential exists for corrosion from moist hydrogen sulfide and cyanides. When feedstocks are from the delayed coker or the TCC, corrosion from hydrogen sulfide and deposits in the high pressure sections of gas compressors from ammonium compounds is possible.
- *Health*: Because these are closed processes, exposures are expected to be minimal under normal operating conditions. There is a potential for exposures to amine compounds such as monoethanolamine (MEA), diethanolamine (DEA) and methyld-iethanolamine (MDEA) and hydrocarbons. Safe work practices and/ or appropriate personal protective equipment may be needed for exposures to chemicals and other hazards such as noise and heat, and during process sampling, inspection, maintenance, and turnaround activities.

AMINE PLANTS

- *Description*: Amine plants remove acid contaminants from sour gas and hydrocarbon streams. In amine plants, gas and liquid hydrocarbon streams containing carbon dioxide and/or hydrogen sulfide are charged to a gas absorption tower or liquid contactor where the acid contaminants are absorbed by counterflowing amine solutions (*i.e.*, MEA, DEA, MDEA). The stripped gas or liquid is removed overhead, and the amine is sent to a regenerator. In the regenerator, the acidic components are stripped by heat and reboiling action and disposed of, and the amine is recycled.

Health and Safety Considerations

- *Fire Protection and Prevention*: The potential for fire exists where a spill or leak could reach a source of ignition.
- *Safety*: To minimize corrosion, proper operating practices should be established and regenerator bottom and reboiler temperatures controlled. Oxygen should be kept out of the system to prevent amine oxidation.
- *Health*: Because this is a closed process, exposures are expected to be minimal during normal operations. There is potential for exposure to amine compounds (*i.e.* monoethanolamine, diethanolamine, methyldiethanolamine), hydrogen sulfide and carbon dioxide. Safe work practices and/or appropriate personal protective equipment may be needed for exposures to chemicals and other hazards such as noise and heat, and during process sampling, inspection, maintenance and turnaround activities.

SATURATE GAS PLANTS

Description

Saturate (sat) gas plants separate refinery gas components including butanes for alkylation, pentanes for gasoline blending, LPG's for fuel, and ethane for petrochemicals. Because sat gas processes depend on the feedstock and product demand, each refinery uses different systems, usually absorption-fractionation or straight fractionation. In absorption-fractionation, gases and liquids from various refinery units are fed to an absorber-deethanizer where C2 and lighter fractions are separated from heavier fractions by lean oil absorption and removed for use as fuel gas or petrochemical feed. The heavier fractions are stripped and sent to a debutanizer, and the lean oil is recycled back to the absorber-deethanizer. C3/C4 is separated from pentanes in the debutanizer, scrubbed to remove hydrogen sulfide, and fed to a splitter where propane and butane are separated. In fractionation sat gas plants, the absorption stage is eliminated.

Health and Safety Considerations

- *Fire Protection and Prevention*: There is potential for fire if a leak or release reaches a source of ignition such as the unit reboiler.
- *Safety*: Corrosion could occur from the presence of hydrogen sulfide, carbon dioxide, and other compounds as a result of prior treating. Streams containing ammonia should be dried before processing. Antifouling additives may be used in absorption oil to protect heat exchangers. Corrosion inhibitors may be used to control corrosion in overhead systems.
- *Health*: Because this is a closed process, exposures are expected to be minimal during normal operations. There is potential for exposure to hydrogen sulfide, carbon dioxide, and other products such as diethanolamine or sodium hydroxide carried over from prior treating. Safe work practices and/or appropriate personal protective equipment may be needed for exposures to chemicals and other hazards such as noise and heat, and during process sampling, inspection, maintenance, and turnaround activities.

ASPHALT PRODUCTION

Description

- Asphalt is a portion of the residual fraction that remains after primary distillation operations. It is further processed to impart characteristics required by its final use. In vacuum distillation, generally used to produce road-tar asphalt, the residual is heated to about 750° F and charged to a column where vacuum is applied to prevent cracking.

- Asphalt for roofing materials is produced by air blowing. Residual is heated in a pipe still almost to its flash point and charged to a blowing tower where hot air is injected for a predetermined time. The dehydrogenization of the asphalt forms hydrogen sulfide, and the oxidation creates sulfur dioxide. Steam, used to blanket the top of the tower to entrain the various contaminants, is then passed through a scrubber to condense the hydrocarbons.
- A third process used to produce asphalt is solvent deasphalting. In this extraction process, which uses propane (or hexane) as a solvent, heavy oil fractions are separated to produce heavy lubricating oil, catalytic cracking feedstock, and asphalt. Feedstock and liquid propane are pumped to an extraction tower at precisely controlled mixtures, temperatures (150°-250° F), and pressures of 350-600 psi. Separation occurs in a rotating disc contactor, based on differences in solubility. The products are then evaporated and steam stripped to recover the propane, which is recycled. Deasphalting also removes some sulfur and nitrogen compounds, metals, carbon residues, and paraffins from the feedstock.

Table. Solvent Deasphalting Process

Feedstock	From	Process	Typical Products	To
Residual, reduced crude	Atmospheric tower & Vacuum tower	Treatment	Heavy lube oil	Treating or lube blending
			Asphalt	Storage of shipping
			Deasphalted oil	Hydrotreat & catalytic cracker
			Propane	Recycle

Health and Safety Considerations

- *Fire Protection and Prevention*: The potential for a fire exists if a product leak or release contacts a source of ignition such as the process heater. Condensed steam from the various asphalt and deasphalting processes will contain trace amounts of hydrocarbons. Any disruption of the vacuum can result in the entry of atmospheric air and subsequent fire. In addition, raising the temperature of the vacuum tower bottom to improve efficiency can generate methane by thermal cracking. This can create vapors in asphalt storage tanks that are not detectable by flash testing but are high enough to be flammable.
- *Safety*: Deasphalting requires exact temperature and pressure control. In addition, moisture, excess solvent, or a drop in operating temperature may cause foaming, which affects the product temperature control and may create an upset.
- *Health*: Because these are closed processes, exposures are expected to be minimal during normal operations. Should a spill or release occur, there is a potential for exposure to residuals and asphalt. Air blowing can create some polynuclear aromatics. Condensed steam from the air-blowing asphalt process may also contain contaminants. The potential for exposure to hydrogen sulfide and sulfur dioxide exists in the production of asphalt. Safe work practices and/or appropriate personal protective equipment may be needed for exposures to chemicals and other hazards such as noise and heat, and during process sampling, inspection, maintenance, and turnaround activities.

HYDROGEN PRODUCTION

Description

- High-purity hydrogen (95%-99%) is required for hydrodesulfurization, hydrogenation, hydro-cracking, and petrochemical processes. Hydrogen, produced as a by-product of refinery processes (principally hydrogen recovery from catalytic reformer product gases), often is not enough to meet the total refinery requirements, necessitating the manufacturing of additional hydrogen or obtaining supply from external sources.

- In steam-methane reforming, desulfurized gases are mixed with superheated steam (1,100°-1,600° F) and reformed in tubes containing a nickel base catalyst. The reformed gas, which consists of steam, hydrogen, carbon monoxide, and carbon dioxide, is cooled and passed through converters containing an iron catalyst where the carbon monoxide reacts with steam to form carbon dioxide and more hydrogen. The carbon dioxide is removed by amine washing. Any remaining carbon monoxide in the product stream is converted to methane.
- Steam-naphtha reforming is a continuous process for the production of hydrogen from liquid hydrocarbons and is, in fact, similar to steam-methane reforming. A variety of naphthas in the gasoline boiling range may be employed, including fuel containing up to 35% aromatics. Following pretreatment to remove sulfur compounds, the feedstock is mixed with steam and taken to the reforming furnace (1,250°-1,500° F) where hydrogen is produced.

Table. Steam Reforming Process.

Feedstock	From	Process	Typical products	To
Desufurized refinery gas	Various treatment units	Decomposition	Hydrogen	Processing
			Carbon dioxide	Atmosphere
			Carbon monoxide	Methane

Health and Safety Considerations

- *Fire Protection and Prevention*: The possibility of fire exists should a leak or release occur and reach an ignition source.
- *Safety*: The potential exists for burns from hot gases and superheated steam should a release occur. Inspections and testing should be considered where the possibility exists for valve failure due to contaminants in the hydrogen. Carryover from caustic scrubbers should be controlled to prevent corrosion in preheaters. Chlorides from the feedstock or steam system should be prevented from entering reformer tubes and contaminating the catalyst.
- *Health*: Because these are closed processes, exposures are expected to be minimal during normal operating conditions. There is a potential for exposure to excess hydrogen, carbon monoxide, and/or carbon dioxide. Condensate can be contaminated by process materials such as caustics and amine compounds, with resultant exposures. Depending on the specific process used, safe work practices and/or appropriate personal protective equipment may be needed for exposures to chemicals and other hazards such as noise and heat, and during process sampling, inspection, maintenance, and turnaround activities.

BLENDING

Description

Blending is the physical mixture of a number of different liquid hydrocarbons to produce a finished product with certain desired characteristics. Products can be blended in-line through a manifold system, or batch blended in tanks and vessels. In-line blending of gasoline, distillates, jet fuel, and kerosene is accomplished by injecting proportionate amounts of each component into the main stream where turbulence promotes thorough mixing.

Additives including octane enhancers, metal deactivators, anti-oxidants, anti-knock agents, gum and rust inhibitors, detergents, etc. are added during and/or after blending to provide specific properties not inherent in hydrocarbons.

Health and Safety Considerations

- *Fire Prevention and Protection*: Ignition sources in the area need to be controlled in the event of a leak or release.
- *Health*: Safe work practices and/or appropriate personal protective equipment may be needed for exposures to chemicals and other hazards such as noise and heat; when handling additives; and during inspection, maintenance, and turnaround activities.

LUBRICANT, WAX, AND GREASE MANUFACTURING PROCESSES

Description

Lubricating oils and waxes are refined from the residual fractions of atmospheric and vacuum distillation. The primary objective of the various lubricating oil refinery processes is to remove asphalts, sulfonated aromatics, and paraffinic and isoparaffinic waxes from residual fractions. reduced crude from the vacuum unit is deasphalted and combined with straight-run lubricating oil feedstock, preheated, and solvent-extracted (usually with phenol or furfural) to produce raffinate.

WAX MANUFACTURING PROCESS

Raffinate from the extraction unit contains a considerable amount of wax that must be removed by solvent extraction and crystallization. The raffinate is mixed with a solvent (propane) and precooled in heat exchangers. The crystallization temperature is attained by the evaporation of propane in the chiller and filter feed tanks. The wax is continuously removed by filters and cold solvent-washed to recover retained oil. The solvent is recovered from the oil by flashing and steam stripping.

The wax is then heated with hot solvent, chilled, filtered, and given a final wash to remove all oil.

LUBRICATING OIL PROCESS

The dewaxed raffinate is blended with other distillate fractions and further treated for viscosity index, colour, stability, carbon residue, sulfur, additive response, and oxidation stability in extremely selective extraction processes using solvents (furfural, phenol, etc.).

Phenol is then separated from the treated oil and recycled. The treated lube-oil base stocks are then mixed and/or compounded with additives to meet the required physical and chemical characteristics of motor oils, industrial lubricants, and metal working oils.

Table. Lubricating Oil and Wax Manufacturing Processes

Feedstock	From	Process	Typical Products	To
Lube feedstock and additives	Vacuum tower, solvent dewaxing, hydrotreating solvent extraction, etc.	Treatment	Dewaxed raffinate	Lube blend or compound, grease compounding
			Wax	Storage or shipping

GREASE COMPOUNDING

Grease is made by blending metallic soaps (salts of long-chained fatty acids) and additives into a lubricating oil medium at temperatures of 400°-600° F. Grease may be either batch-produced or continuously compounded. The characteristics of the grease depend to a great extent on the metallic element (calcium, sodium, aluminum, lithium, etc.) in the soap and the additives used.

SAFETY AND HEALTH CONSIDERATIONS

- *Fire Protection and Prevention*: The potential for fire exists if a product or vapour leak or release in the lube blending and wax processing areas reaches a source of ignition. Storage of finished products, both bulk and packaged, should be in accordance with recognized practices. While the potential for fire is reduced in lube oil blending, care must be taken when making metal-working oils and compounding greases due to the use of higher blending and compounding temperatures and lower flash point products.

- *Safety*: Control of treater temperature is important as phenol can cause corrosion above 400° F. Batch and in-line blending operations require strict controls to maintain desired product quality. Spills should be cleaned and leaks repaired to avoid slips and falls. Additives in drums and bags need to be handled properly to avoid strain. Wax can clog sewer or oil drainage systems and interfere with wastewater treatment.
- *Health*: When blending, sampling, and compounding, personal protection from steam, dusts, mists, vapors, metallic salts, and other additives is appropriate. Skin contact with any formulated grease or lubricant should be avoided. Safe work practices and/or appropriate personal protection may be needed for exposures to chemicals and other hazards such as noise and heat; during inspection, maintenance, and turnaround activities; and while sampling and handling hydrocarbons and chemicals during the production of lubricating oil and wax.

HEAT EXCHANGERS, COOLERS, AND PROCESS HEATERS

Heating Operations

Process heaters and heat exchangers preheat feedstock in distillation towers and in refinery processes to reaction temperatures. The heaters are usually designed for specific process operations, and most are of cylindrical vertical or box-type designs. The major portion of heat provided to process units comes from fired heaters fueled by refinery or natural gas, distillate, and residual oils. Fired heaters are found on crude and reformer preheaters, coker heaters, and large-column reboilers.

Cooling Operations

Heat also may be removed from some processes by air and water exchangers, fin fans, gas and liquid coolers, and overhead condensers, or by transferring heat to other systems. The basic mechanical vapour-compression refrigeration system, which may serve one or more process units, includes an evaporator, compressor, condenser, controls, and piping. Common coolants are water, alcohol/water mixtures, or various glycol solutions.

Health and Safety Considerations

- *Fire Protection and Prevention*: A means of providing adequate draft or steam purging is required to reduce the chance of explosions when lighting fires in heater furnaces. Specific start-up and emergency procedures are required for each type of unit. If fire impinges on fin fans, failure could occur due to overheating. If flammable product escapes from a heat exchanger or cooler due to a leak, fire could occur.
- *Safety*: Care must be taken to ensure that all pressure is removed from heater tubes before removing header or fitting plugs. Consideration should be given to providing for pressure relief in heat-exchanger piping systems in the event they are blocked off while full of liquid. If controls fail, variations of temperature and pressure could occur on either side of the heat exchanger. If heat exchanger tubes fail and process pressure is greater than heater pressure, product could enter the heater with downstream consequences. If the process pressure is less than heater pressure, the heater stream could enter into the process fluid. If loss of circulation occurs in liquid or gas coolers, increased product temperature could affect downstream operations and require pressure relief.

- *Health*: Because these are closed systems, exposures under normal operating conditions are expected to be minimal. Depending on the fuel, process operation, and unit design, there is a potential for exposure to hydrogen sulfide, carbon monoxide, hydrocarbons, steam boiler feed-water sludge, and water-treatment chemicals. Skin contact should be avoided with boiler blowdown, which may contain phenolic compounds. Safe work practices and/or appropriate personal protective equipment against hazards may be needed during process maintenance, inspection, and turnaround activities and for protection from radiant heat, superheated steam, hot hydrocarbon, and noise exposures.

STEAM GENERATION

Heater and Boiler Operations

Steam is generated in main generation plants, and/or at various process units using heat from flue gas or other sources. Heaters (furnaces) include burners and a combustion air system, the boiler enclosure in which heat transfer takes place, a draft or pressure system to remove flue gas from the furnace, soot blowers, and compressed-air systems that seal openings to prevent the escape of flue gas.

Boilers consist of a number of tubes that carry the water-steam mixture through the furnace for maximum heat transfer. These tubes run between steam-distribution drums at the top of the boiler and water-collecting drums at the bottom of the boiler. Steam flows from the steam drum to the superheater before entering the steam distribution system.

Heater Fuel

- Heaters may use any one or combination of fuels including refinery gas, natural gas, fuel oil, and powdered coal. Refinery off-gas is collected from process units and combined with natural gas and LPG in a fuel-gas balance drum. The balance drum provides constant system pressure, fairly stable Btu-content fuel, and automatic separation of suspended liquids in gas vapors, and it prevents carryover of large slugs of condensate into the distribution system. Fuel oil is typically a mix of refinery crude oil with straight-run and cracked residues and other products. The fuel-oil system delivers fuel to process-unit heaters and steam generators at required temperatures and pressures. The fuel oil is heated to pumping temperature, sucked through a coarse suction strainer, pumped to a temperature-control heater, and then pumped through a fine-mesh strainer before being burned.

- In one example of process-unit heat generation, carbon monoxide boilers recover heat in catalytic cracking units as carbon monoxide in flue gas is burned to complete combustion. In other processes, waste-heat recovery units use heat from the flue gas to make steam.

Steam Distribution

The distribution system consists of valves, fittings, piping, and connections suitable for the pressure of the steam transported. Steam leaves the boilers at the highest pressure required by the process units or electrical generation. The steam pressure is then reduced in turbines that drive process pumps and compressors. Most steam used in the refinery is condensed to water in various types of heat exchangers.

The condensate is reused as boiler feedwater or discharged to wastewater treatment. When refinery steam is also used to drive steam turbine generators to produce electricity, the steam must be produced at much higher pressure than required for process steam. Steam typically is generated by heaters (furnaces) and boilers combined in one unit.

Feedwater

- Feedwater supply is an important part of steam

generation. There must always be as many pounds of water entering the system as there are pounds of steam leaving it. Water used in steam generation must be free of contaminants including minerals and dissolved impurities that can damage the system or affect its operation. Suspended materials such as silt, sewage, and oil, which form scale and sludge, must be coagulated or filtered out of the water. Dissolved gases, particularly carbon dioxide and oxygen, cause boiler corrosion and are removed by deaeration and treatment. Dissolved minerals including metallic salts, calcium, carbonates, etc., that cause scale, corrosion, and turbine blade deposits are treated with lime or soda ash to precipitate them from the water. Recirculated cooling water must also be treated for hydrocarbons and other contaminants.

- Depending on the characteristics of raw boiler feedwater, some or all of the following six stages of treatment will be applicable:
  - Clarification
  - Sedimentation
  - Filtration
  - Ion exchange
  - Deaeration
  - Internal treatment

Health and Safety Considerations

- *Fire Protection and Prevention*: The most potentially hazardous operation in steam generation is heater startup. A flammable mixture of gas and air can build up as a result of loss of flame at one or more burners during light-off. Each type of unit requires specific startup and emergency procedures including purging before lightoff and in the event of misfire or loss of burner flame.
- *Safety*: If feedwater runs low and boilers are dry, the tubes will overheat and fail. Conversely, excess water

will be carried over into the steam distribution system and damage the turbines. Feedwater must be free of contaminants that could affect operations. Boilers should have continuous or intermittent blowdown systems to remove water from steam drums and limit buildup of scale on turbine blades and superheater tubes. Care must be taken not to overheat the superheater during startup and shut-down. Alternate fuel sources should be provided in the event of loss of gas due to refinery unit shutdown or emergency. Knockout pots provided at process units remove liquids from fuel gas before burning.

- *Health*: Safe work practices and/or appropriate personal protective equipment may be needed for potential exposures to feedwater chemicals, steam, hot water, radiant heat, and noise, and during process sampling, inspection, maintenance, and turnaround activities.

PRESSURE-RELIEF AND FLARE SYSTEMS

Pressure-Relief Systems

Pressure-relief systems control vapors and liquids that are released by pressure-relieving devices and blow-downs. Pressure relief is an automatic, planned release when operating pressure reaches a predetermined level. Blowdown normally refers to the intentional release of material, such as blowdowns from process unit startups, furnace blowdowns, shutdowns, and emergencies.

Vapour depressuring is the rapid removal of vapors from pressure vessels in case of fire. This may be accomplished by the use of a rupture disc, usually set at a higher pressure than the relief valve.

Safety Relief Valve Operations

Safety relief valves, used for air, steam, and gas as well as for vapour and liquid, allow the valve to open in proportion to the increase in pressure over the normal operating pressure.

Safety valves designed primarily to release high volumes of steam usually pop open to full capacity. The overpressure needed to open liquid-relief valves where large-volume discharge is not required increases as the valve lifts due to increased spring resistance.

Pilot-operated safety relief valves, with up to six times the capacity of normal relief valves, are used where tighter sealing and larger volume discharges are required. Nonvolatile liquids are usually pumped to oil-water separation and recovery systems, and volatile liquids are sent to units operating at a lower pressure.

Flare Systems

A typical closed pressure release and flare system includes relief valves and lines from process units for collection of discharges, knockout drums to separate vapors and liquids, seals, and/or purge gas for flashback protection, and a flare and igniter system which combusts vapors when discharging directly to the atmosphere is not permitted. Steam may be injected into the flare tip to reduce visible smoke.

WASTEWATER TREATMENT

Description

Wastewater treatment is used for process, run-off, and sewerage water prior to discharge or recycling. Wastewater typically contains hydrocarbons, dissolved materials, suspended solids, phenols, ammonia, sulfides, and other compounds. Wastewater includes condensed steam, stripping water, spent caustic solutions, cooling tower and boiler blowdown, wash water, alkaline and acid waste neutralization water, and other process-associated water.

Pretreatment Operations

Pretreatment is the separation of hydrocarbons and solids from wastewater. API separators, interceptor plates, and settling ponds remove suspended hydrocarbons, oily sludge, and solids by gravity separation, skimming, and filtration. Some oil-in-water emulsions must be heated to assist in separating the oil and water. Gravity separation depends on the specific gravity differences between water and immiscible oil globules and allows free oil to be skimmed off the surface of the wastewater. Acidic wastewater is neutralized using ammonia, lime, or soda ash. Alkaline wastewater is treated with sulfuric acid, hydrochloric acid, carbon dioxide-rich flue gas, or sulfur.

Secondary Treatment Operations

After pretreatment, suspended solids are removed by sedimentation or air flotation. Wastewater with low levels of solids may be screened or filtered. Flocculation agents are sometimes added to help separation. Secondary treatment processes biologically degrade and oxidize soluble organic matter by the use of activated sludge, unaerated or aerated lagoons, trickling filter methods, or anaerobic treatments. Materials with high adsorption characteristics are used in fixed-bed filters or added to the wastewater to form a slurry which is removed by sedimentation or filtration. Additional treatment methods are used to remove oils and chemicals from wastewater. Stripping is used on wastewater containing sulfides and/or ammonia, and solvent extraction is used to remove phenols.

Tertiary Treatment Operations

Tertiary treatments remove specific pollutants to meet regulatory discharge requirements. These treatments include chlorination, ozonation, ion exchange, reverse osmosis, activated carbon adsorption, etc. Compressed oxygen is diffused into wastewater streams to oxidize certain chemicals or to satisfy regulatory oxygen-content requirements. Wastewater that is to be recycled may require cooling to remove heat and/or oxidation by spraying or air stripping to remove any remaining phenols, nitrates, and ammonia.

Health and Safety Considerations

- *Fire Protection and Prevention*: The potential for fire

exists if vapors from wastewater containing hydrocarbons reach a source of ignition during treatment.

- *Health*: Safe work practices and/or appropriate personal protective equipment may be needed for exposures to chemicals and waste products during process sampling, inspection, maintenance, and turnaround activities as well as to noise, gases, and heat.

COOLING TOWERS

Description

Cooling towers remove heat from process water by evaporation and latent heat transfer between hot water and air. The two types of towers are crossflow and counterflow. Crossflow towers introduce the airflow at right angles to the water flow throughout the structure. In counterflow cooling towers, hot process water is pumped to the uppermost plenum and allowed to fall through the tower.

Numerous slats or spray nozzles located throughout the length of the tower disperse the water and help in cooling. Air enters at the tower bottom and flows upward against the water. When the fans or blowers are at the air inlet, the air is considered to be forced draft. Induced draft is when the fans are at the air outlet.

Cooling Water

Recirculated cooling water must be treated to remove impurities and dissolved hydrocarbons. Because the water is saturated with oxygen from being cooled with air, the chances for corrosion are increased. One means of corrosion prevention is the addition of a material to the cooling water that forms a protective film on pipes and other metal surfaces.

Health and Safety Considerations

- *Fire Prevention and Protection*: When cooling water is

contaminated by hydrocarbons, flammable vapors can be evaporated into the discharge air. If a source of ignition is present, or if lightning occurs, a fire may start. A potential fire hazard also exists where there are relatively dry areas in induced-draft cooling towers of combustible construction.

- *Safety*: Loss of power to cooling tower fans or water pumps could have serious consequences in the operation of the refinery. Impurities in cooling water can corrode and foul pipes and heat exchangers, scale from dissolved salts can deposit on pipes, and wooden cooling towers can be damaged by microorganisms.
- *Health*: Cooling-tower water can be contaminated by process materials and by-products including sulfur dioxide, hydrogen sulfide, and carbon dioxide, with resultant exposures. Safe work practices and/or appropriate personal protective equipment may be needed during process sampling, inspection, maintenance, and turnaround activities; and for exposure to hazards such as those related to noise, water-treatment chemicals, and hydrogen sulfide when wastewater is treated in conjunction with cooling towers.

ELECTRIC POWER

Description

Refineries may receive electricity from outside sources or produce their own power with generators driven by steam turbines or gas engines. Electrical substations receive power from the utility or power plant for distribution throughout the facility. They are usually located in nonclassified areas, away from sources of vapour or cooling-tower water spray. Transformers, circuit breakers, and feed-circuit switches are usually located in substations.

Substations feed power to distribution stations within the process unit areas. Distribution stations can be located in classified areas, providing that classification requirements are met. Distribution stations usually have a liquid-filled transformer and an oil-filled or air-break disconnect device.

Health and Safety Considerations

- *Fire Protection and Prevention*: Generators that are not properly classified and are located too close to process units may be a source of ignition should a spill or release occur.
- *Safety*: Normal electrical safety precautions including dry footing, high-voltage warning signs, and guarding must be taken to protect against electrocution. Lockout/tagout and other appropriate safe work practices must be established to prevent energization while work is being performed on high-voltage electrical equipment.
- *Health*: Safe work practices and/or the use of appropriate personal protective equipment may be needed for exposures to noise, for exposure to hazards during inspection and maintenance activities, and when working around transformers and switches that may contain a dielectric fluid which requires special handling precautions.

GAS AND AIR COMPRESSORS

Description

Both reciprocating and centrifugal compressors are used throughout the refinery for gas and compressed air. Air compressor systems include compressors, coolers, air receivers, air dryers, controls, and distribution piping. Blowers are used to provide air to certain processes.

Plant air is provided for the operation of air-powered tools, catalyst regeneration, process heaters, steam-air decoking, sour-water oxidation, gasoline sweetening, asphalt blowing, and other uses. Instrument air is provided for use in pneumatic instruments and controls, air motors and purge connections.

Health and Safety Considerations

- *Fire Protection and Prevention*: Air compressors should be located so that the suction does not take in flammable vapors or corrosive gases. There is a potential for fire should a leak occur in gas compressors.
- *Safety*: Knockout drums are needed to prevent liquid surges from entering gas compressors. If gases are contaminated with solid materials, strainers are needed. Failure of automatic compressor controls will affect processes. If maximum pressure could potentially be greater than compressor or process-equipment design pressure, pressure relief should be provided. Guarding is needed for exposed moving parts on compressors. Compressor buildings should be properly electrically classified, and provisions should be made for proper ventilation. Where plant air is used to back up instrument air, interconnections must be upstream of the instrument air drying system to prevent contamination of instruments with moisture. Alternate sources of instrument air supply, such as use of nitrogen, may be needed in the event of power outages or compressor failure.
- *Health*: Safe work practices and/or appropriate personal protective equipment may be needed for exposure to hazards such as noise and during inspection and maintenance activities. The use of appropriate safeguards must be considered so that plant and instrument air is not used for breathing or pressuring potable water systems.

9

Gas-Crude Oil Sample

DESCRIPTION

The laboratory procedure for sample analysis generally follows this procedure:

- Determine the mole fraction composition of the subsurface sample, or of the individual gas and oil surface samples.
- Determine shrinkage for surface separator samples.
- Recombine the surface oil and gas samples according to test data.
- Perform a relative total volume or PV test.
- Perform a differential liberation test at reservoir temperature in the test cell.
- Perform a flash separation test at various separation pressures and temperatures.
- Determine the fluid viscosity over a range of pressures at reservoir temperature.

The first page or two of a report generally is a cover letter which includes a statement of the field quality check on the samples and of the tests performed. The bubble-point pressure at reservoir temperature, the total solution gas and relative volume factor for differential liberation, the range of oil viscosities, and the optimum separator pressure are usually summarized in the text. Following this is a listing of formation characteristics, well characteristics, and sampling conditions. All of these data are obtained before or during the sampling procedure. Some testing laboratories may include more extensive amounts of the field test data and conditioning procedures.

Most important are the reservoir pressure, gas and oil rates, and reservoir temperature, although all of the data are helpful and possibly critical in interpreting the test results.

HYDROCARBON COMPOSITIONAL ANALYSTS

Next in the report we will find the compositional analyses of the reservoir fluid. If a surface recombination sample has been taken, the compositionof the individual separator gas and liquid samples is listed along with the composition of the recombined reservoir fluid. The basis of recombination is always given, as well as the molecular weight and density of the heptanes-plus (C7+) fraction and the properties of separator gas and liquid.

Composition may be determined by fractionation in a low or high temperature fractionating column, by mass spectrometer, or, more commonly, by gas chromatography. A chromatograph, Gas chromatograph used for compositional analysis of hydrocarbon mixtures) separates the components according to boiling point, in a special column.

Fig. 9.1

Liquids are analysed by capillary gas chromatography up to a certain molecular weight (C35), and the heavier fraction is characterized by molecular weight and density. Molecular weight measurements are performed on the heavier fraction by cryoscopy. In this technique the freezing point of a mixture of benzene and the oil's heavy ends is measured and compared to the freezing point of pure benzene. This allows a determination of the heavy fraction's molecular weight and density. As mentioned earlier, the recombination of the gas and oil samples must be made based on the separator GOR expressed with respect to separator liquid. If a separator-stock tank oil shrinkage factor was not measured in the field, it must be determined in the laboratory by recreating separator and stock tank conditions and measuring the difference in volumes of the separator and stock tank liquid. Only on this basis can the samples be recombined and the fluid composition analysed.

RELATIVE TOTAL VOLUME TEST

After the representative fluid sample is transferred into the test cell, usually a high-pressure, high-temperature, mercury cell (High-pressure, high-temperature mercury test cell for relative volume determinations.

Fig.9.2

Viewing mirror allows view glass to be pointed away from operator), the flash separation procedure shown in Figure 9.3 may be carried out.

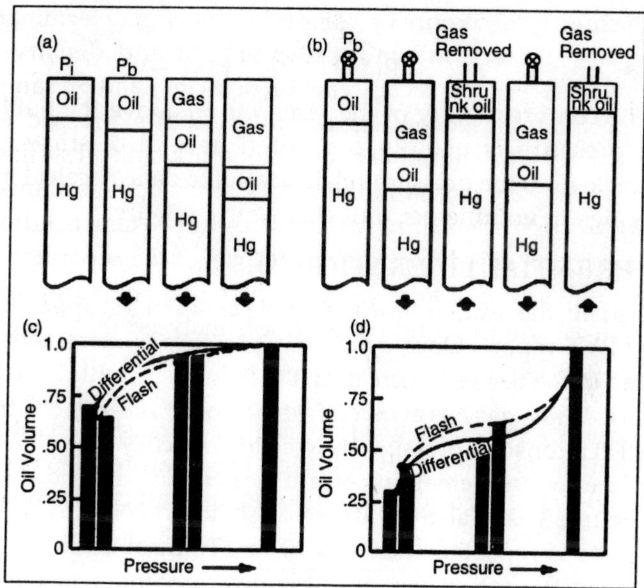

Fig.9.3

(Schematic of (a) flash liberation and (b) differential liberation. The degree to which oil volume is effected by the separation process is dependent on the composition of the oil. In the case of a low shrinkage oil (c), differential liberation provides for a larger volume of stock tank oil. A high shrinkage oil (d) is affected differently.) The pressure on the sample is decreased from above reservoir pressure by withdrawing mercury at reservoir temperature.

The pressure and volume changes allow the plotting of a graph to determine the bubble point. Once gas begins to appear, the sample is brought to equilibrium after each volume change by rocking the cell. The volume is expressed relative to the volume at the bubble-point or saturation pressure.

No material is removed from the cell during this procedure. Usually, the data entered in the PVT report is "smoothed" to account for the errors in reading very small changes in volume. This smoothing is done by fitting the total relative volume data to a "Y" function curve:

$$Y = \frac{pb - p}{p[(Vt/Vb) - 1]}$$

The compressibility of the reservoir oil above the bubble-point pressure is usually presented here also, since it is obtainable from the pressure-relative volume data by calculating the change in volume per unit change in pressure.

DIFFERENTIAL LIBERATION TEST

Usually the results of a differential liberation test appear next in our PVT report. Once again, the pressure is incrementally lowered in the test cell, increasing the volume by withdrawal of mercury. When gas has evolved from the oil, it is displaced from the cell at a constant pressure by injecting mercury. The volumes of the free gas and remaining oil are measured at cell conditions, and the free gas is also measured at standard conditions. This procedure is repeated (perhaps 10 or 12 times) until only oil remains in the cell at reservoir temperature and atmospheric pressure. The residual oil is then cooled to standard temperature in order to measure its change in volume. This residual oil is not the same as stock tank oil, because the differential process of gas liberation is not equivalent to flash liberation. Differential liberation is carried out at reservoir temperature, thus the residual oil will have more of the lighter ends "cooked off."

This results in the gravity of the residual oil being lower than that of an oil resulting from a flash liberation carried out at separator temperature. In the differential liberation test, material is removed from the system, changing the overall composition of the hydrocarbon mixture in the cell. The relative oil volume data presented in the PVT report are given relative to a volume of residual oil at standard conditions. The solution gas-oil ratio is also presented, based on the volume of gas liberated in each incremental pressure drop and the total gas released. The compressibility factor, formation volume factor, and gas gravity are determined for the gas released during each pressure decrement, and presented in the report. The oil density also is reported here. The relative oil volume data measured so far are not equivalent to the oil formation volume factor we wish to obtain for use in material balance and reserve calculations. In order to adjust the differential liberation data to obtain B0, we need to simulate the flash liberation process that takes place in the separator and results in stock tank oil. This is done in the flash separation test.

FLASH SEPARATION TESTS

In this test, part of the reservoir fluid is ejected from the test cell at reservoir temperature and saturation pressure into a small-scale stage-separation system. The separation pressures and temperatures are carefully controlled and generally determined by the engineer requesting the test. The volume of gas liberated at each stage and the volume of remaining liquid are measured. Normally the test is carried out for two stages of separation. The first-stage separator pressure is generally varied to include at least four possible separator pressures at ambient temperature, and the second stage is generally carried out under stock tank conditions of o psig (1 atmosphere) at ambient temperature. The data presented for this test include the GOR in gas volumes per separator barrel and per stock tank barrel. The formation volume factor given here is the volume of saturated oil at bubble-point pressure and reservoir temperature per volume of stock tank oil at standard conditions.

The gravity of the oil and the gravity of the flashed gas are also reported for each separator pressure. These data may be plotted as a function of separator pressure, Formation volume factor and oil gravity versus separator pressure) to reveal the optimum separator pressure. The optimum pressure is that which maximizes stock tank oil gravity and volume. Assuming that we will be able to install a separator that operates at this optimum pressure, we can use the flash separation data from the stage separation closest to this optimum pressure to adjust our earlier differential data, thus obtaining B0 for the proposed field conditions. This is done by multiplying each differential liberation volume factor by the ratio of the flash separation volume factor and the differential liberation volume factor at the bubble point. In Figure 9.4.

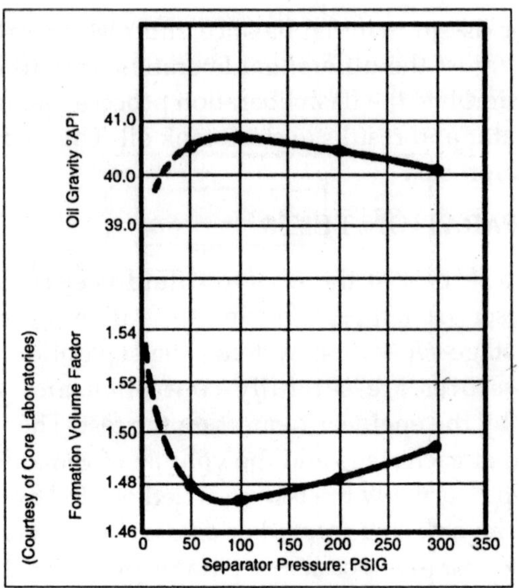
(Courtesy of Core Laboratories)
Oil Gravity °API
41.0
40.0
39.0
Formation Volume Factor
1.54
1.52
1.50
1.48
1.46
0
50
100
150
200
250
300
350
Separator Pressure: PSIG

Fig.9.4

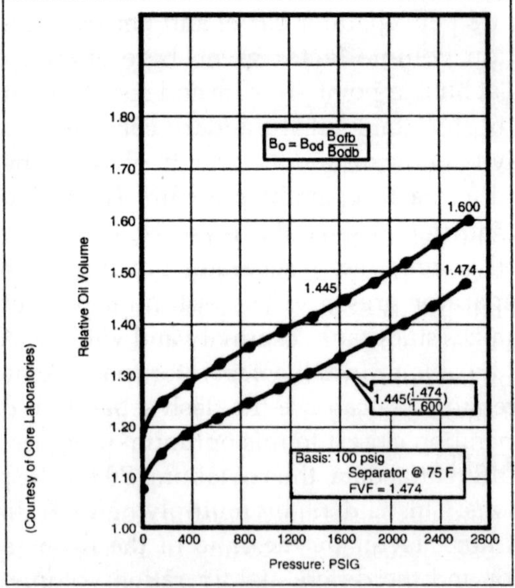
(Courtesy of Core Laboratories)
Relative Oil Volume
1.80
1.70
1.60
1.50
1.40
1.30
1.20
1.10
1.00
$B_o = B_{od} \frac{B_{ofb}}{B_{odb}}$
1.600
1.474
1.445
$1.445(\frac{1.474}{1.600})$
Basis: 100 psig
Separator @ 75 F
FVF = 1.474
0
400
800
1200
1600
2000
2400
2800
Pressure: PSIG

Fig.9.5

Similarly, the solution GOR curve must be adjusted as shown in Figure 9.5 (Adjustment of differential solution GOR curve to separator conditions).

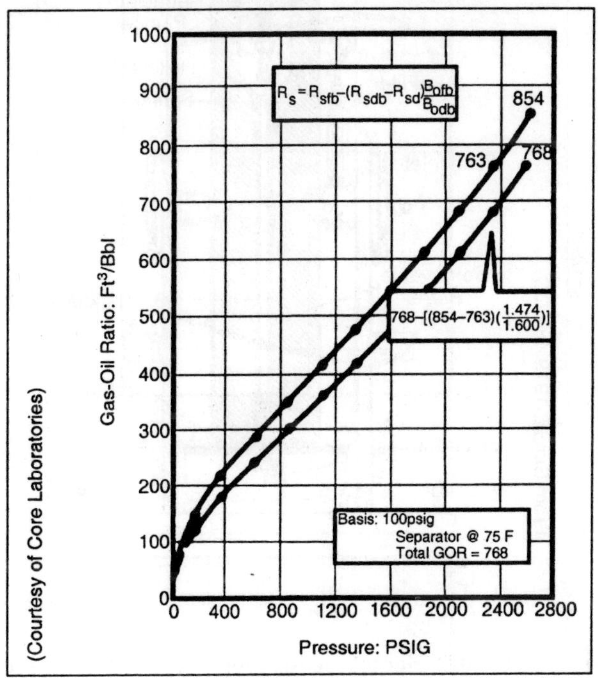

Fig.9.6

The Rs and B0 values generated in this manner may be used to predict reservoir performance when surface separation equipment performs as expected. Some laboratories present adjusted formation volume factor and solution GORs in their reports based on the optimal separator pressure. Other labs leave the selection of an optimal separator pressure up to the engineer.

FLUID VISCOSITY

Reservoir fluid viscosity is measured at reservoir conditions using a rolling-ball viscosimeter (Rolling-ball viscosimeter used for determining fluid viscosity at high pressures and temperatures).

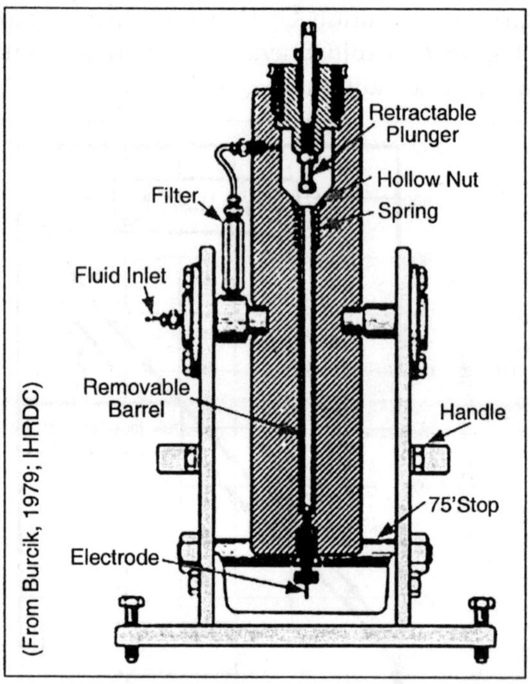

Fig.9.7

This instrument electronically measures the time required for a steel ball to roll a given distance through a tube filled with the fluid to be tested. The clearance between the tube and the ball is varied according to the general magnitude of the fluid viscosity. The viscosity of the oil above the bubble point and below the bubble point is measured in this manner. Gas viscosities are generally determined from the gas analysis using correlations.

By performing a compositional analysis on the gas liberated at any point during differential liberation or flash separation tests, the gas viscosity can be determined. When measured, gas viscosities are usually measured by determining the pressure drop for flow through a capillary tube. The PVT report generally includes plots of all the important tabular data, along with a plot of the subsurface pressure survey if a subsurface sample was taken. Although not part of a typical PVT report, the determination of equilibrium ratios for a reservoir fluid may be done at special request. A test cell is charged with a reservoir fluid sample and the sample is flashed by withdrawing mercury and dropping the pressure. Samples of the equilibrium oil and gas are individually withdrawn and analysed using gas chromatography. The equilibrium ratio at the particular pressure and reservoir temperature can be determined from the analyses. The process is repeated with a fresh sample to a lower pressure than the previous one. This is done several more times over a desired range of pressures.

MATERIAL BALANCE METHOD FOR GAS SYSTEMS

GAS MATERIAL BALANCES AND STRAIGHT LINE METHODS

Gas-in-Place with Water Influx

In prior discussion of MBE, we multipled Gp by Bg, the formation volume factor to obtain a reservoir volume in either cubic ft or bbls. Remember the constant 0.02829 was used to calculate Bg in cubic ft, and 0.00504 for barrels.

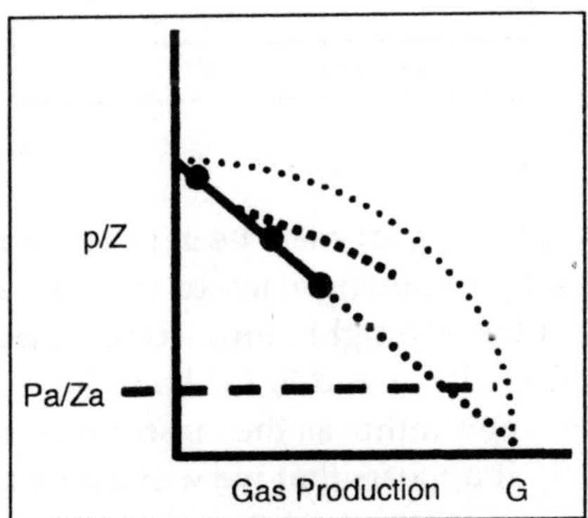

Fig. 9.8

Recovery of a dry gas with adequate production history and no water influx, can be determined by the P/Z verses Gp plot. Overestimation of OGIP and ultimate recovery is possible in the case of a geopressured formation or water influx (purple line). The pressure support from an aquifer may not be apparent at first. Assuming no water influx when one is present can result in over estimation of both gas in place and ultimate recovery.

In the plot 9.8, Pa is the abandonment pressure, with corresponding Za factor. Frequently, the abandonment point is based more on economics, which may relate to the gas production from several fields. From straight line MB methods, we intend to calculate GIIP and cumulative water influx, We. This is a function of aquifer's geometry and transmissibility T = Kh. Aquifers may initially provide complete support or partial support depending on these characteristics.

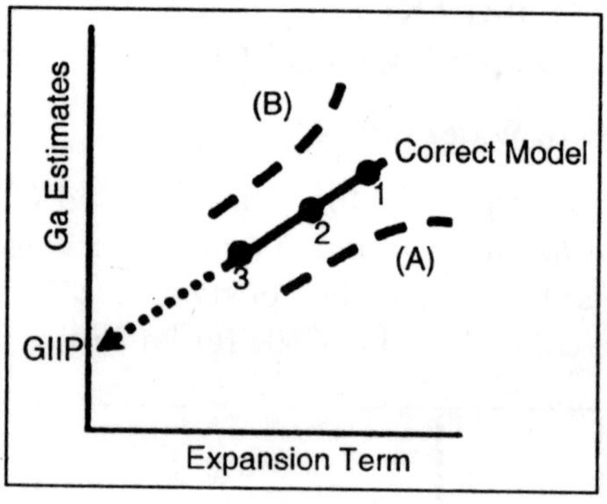

Fig. 9.9

The next graph has purposely been left unlabeled. This of course, is the straight line approach to material balance, and the correct model has the right cumulative water influx, We. The points are labeled 1, 2 and 3, to show the time sequence. You can think of water influx as the opposite of production.

With Line (B), it appears that we will over estimate gas in place. Therefore, we must be using an We that is too high or too low? Think for a moment before proceeding. Congratulations!

We is too low. Just as the gas-in-place is too high when we neglect completely neglect water influx in the P/Z method, GIIP is overestimated when we underestimate it. The straight line equation for gas reservoir with water influx is:

$$\frac{G_F B_g + W_F B_W}{\left(B_g - B_{gi}\right)} = \frac{C \cdot f(p,t)}{B_g - B_{g,}} + G$$

in the Y = mX+b format, where the line has slope C and intercept G.

Steady-State Model for Aquifer

The steady-state model means that the aquifer keeps up with production. Perhaps initially, the field declines in pressure and rates drop off. At this point, the total voidage in the reservoir equals the influx from the aquifer and the pressure stabilizes. In the steady state model, water influx rate (that is the derivative of We with time), is in proportion to the pressure drop. Under this condition, the water influx rate must equal the gas and water rate, all expressed in reservoir units. $\Delta W_e/\Delta t = B_g \Delta G_F/\Delta t + B_w \Delta W_F/\Delta t$ and $C = \Delta W_e/(p_i - p_s)$ where p_i = initial pressure and p_3 = stabilized pressure

Unsteady-State Aquifer

Similar to the aquifer model are the diffusivity constant, which divides storage terms (porosity, total compressibility) by transmissibility (permeability/viscosity). The HE model was later extended by Coats to include bottom water drive. Vertical permeability was added to the model.

Other Models

Will add a link to this topic soon. Note that the geopressured reservoir appears at first to be an enormous field by conventional P/Z techniques.

Final Comments

Theoretical models of aquifers are the same for gas and oil bearing reservoirs, however oil reservoirs add complications due to gas evolution from oil. The straight line methodology attempts to identify the GIIP and C (aquifer constant) that would explain the pressure decline in the field. The alternative is history matching with reservoir simulation. With increases in the speed of these programmes, the use of MB straight line methods are in decline, at least in my experience. The modern approach to simulation is to model flow up the wellbore and provide a complete forecast including well head pressures.

Avoid material balance methods when tank like assumptions will not hold. However, the MB method is far from dead. The simplified model may help assist or give further support to a history match of a gas-aquifer system. Interestingly, material balance methodology is the preferred method for coalbed methane analysis for gas in place calculation.

Index

A

B

C

D

E

F

G

H

V

W

Y

Z